Security in Fixed and Wireless Networks

Security in Fixed and Wireless Networks

An Introduction to Securing Data Communications

Günter Schäfer
Technische Universität, Berlin, Germany

Translated by Hedwig Jourdan von Schmoeger,
London, UK

WILEY

First published under the title Netzsicherheit. Algorithmische Grundlagen und Protokolle
ISBN: 3-89864-212-7 by dpunkt. verlag GmbH
© dpunkt. verlag GmbH, Heidelberg, Germany, 2003

Other Wiley Editorial Offices

John Wiley & Sons, Inc., 111 River Street, Hoboken, NJ 07030, USA

Jossey-Bass, 989 Market Street, San Francisco, CA 94103-1741, USA

Wiley-VCH Verlag GmbH, Boschstr. 12, D-69469 Weinheim, Germany

John Wiley & Sons Australia Ltd, 33 Park Road, Milton, Queensland 4064, Australia

John Wiley & Sons (Asia) Pte Ltd, 2 Clementi Loop #02-01, Jin Xing Distripark, Singapore 129809

John Wiley & Sons (Canada) Ltd, 22 Worcester Road, Etobicoke, Rexdale, Ontario, Canada M9W 1L1

Wiley also publishes its books in a variety of electronic formats. Some content that appears in print may
not be available in electronic books.

British Library Cataloguing in Publication Data

A catalogue record for this book is available from the British Library

ISBN 0-470-86370-6

Typeset from pdf files supplied by the author.
Printed and bound in Great Britain by Antony Rowe, Chippenham, Wiltshire.
This book is printed on acid-free paper responsibly manufactured from sustainable forestry in which at
least two trees are planted for each one used for paper production.

About the Author

Dr.–Ing. Günter Schäfer studied computer science at the Universität Karlsruhe, Germany, from 1989 to 1994. After his studies he continued there as a member of the scientific staff at the Institute of Telematics. He received his doctorate on the topic Efficient Authentication and Key Management in High-Performance Networks in October 1998. In February 1999 Dr Schäfer took a postdoctoral position at the Ecole Nationale Supérieure des Télécommunications in Paris, France, where he focused on network security and access network performance of third-generation mobile communication networks.

Since August 2000, Dr Schäfer has been at the Technische Universität Berlin, Germany, where he is involved in research and lectures on the subject of telecommunications networks. His main subject areas are network security, mobile communications, and active network technologies.

Günter Schäfer is a member of the Institute of Electrical and Electronics Engineers (IEEE) and the Gesellschaft für Informatik (German Computer Science Society).

Acknowledgements

This book has evolved during my time as a scientific assistant in the department of telecommunication networks at the Technische Universität Berlin, Germany. It is based on my lecture, Network Security, which I have been presenting at the University since the winter semester of 2000/2001.

I therefore particularly want to express my warm gratitude to the head of this department, Professor Adam Wolisz, for the wonderful opportunities he has given me for my work. He has supported my plans to write a textbook on network security from the very beginning.

Dipl.-Ing. Mr Andreas Hess offered to read and edit the entire first draft of my text. I am sincerely grateful to him for his fast turnaround times and numerous helpful suggestions for changes and improvements.

Mrs Hedwig Jourdan von Schmoeger translated the German version of the book into English. She not only had a good grasp of the technical content but also had a knack for dealing with my often rather long German sentences. I want to thank her for the very good working relationship we had.

This gratitude also extends to the editorial staff of dpunkt.verlag and John Wiley & Sons, who were so helpful with both the German and English versions of the book. Their constant support and guidance made my task much easier. I also appreciate the helpful input from the various reviewers who provided useful and constructive comments.

Lastly, I want to thank the students who attended my lectures for their numerous questions and suggestions that gave me many ideas for how to structure this book. The responsibility for any errors that still might appear in this book despite all the help that was available, of course, lies with me. I will, therefore, continue to appreciate any comments or suggestions regarding the content of this book.

Berlin, December 2003

Günter Schäfer
(securitybook@guenterschaefer.de)

This book has an accompying website that contains suport material for lecturers as well as sample chapters.

Please visit http://www.guenterschaefer.de/SecurityBook

Contents

I Foundations of Data Security Technology

II   Network Security

Part I

Foundations of Data Security Technology

1   Introduction

It is now a well-known fact that, despite all the benefits, the digital revolution with its omnipresent networking of information systems also involves some risks. This book looks at a specific category of risks and, in particular, the measures that can be taken to minimise them. The category of risks discussed has evolved as a result of eavesdropping and the manipulation of data transmitted in communication networks and the vulnerability of the communication infrastructure itself.

Mankind very early on recognised the need to protect information that was being transferred or stored, and so the desire to protect information from unauthorised access is probably as old as writing itself. For example, reliable early records on protective measures describe a technique used by the Spartans around 400 BC. The technique entailed writing messages on a leather strip that was wrapped around a stick of a certain diameter. Before the message was delivered, the leather strip was removed from the stick, and a potential attacker who did not have a stick with the same diameter (because he did not know the diameter or anything about the technique) could not read the message. In a sense this was an implementation of the first 'analogue' encryption.

Protecting transmitted data

In the fourth century BC, the Greek Polybius developed a table of bilateral substitution that defined how to encode characters into pairs of symbols and their corresponding reinstatement, thereby specifying the first 'digital' encryption method. Of the Romans we know that they often protected their tactical communication by using simple monoalphabetic substitution methods. The best known one was probably the 'Caesar cipher', named after its creator Julius Caesar, in which each character of the alphabet is shifted upwards by three characters. Thus, 'A' becomes 'D', 'B' becomes 'E', etc.

First substitution ciphers

The Arabs were the first people to develop a basic understanding of the two fundamental principles of *substitution*, i.e., pure character substitution, and *transposition*, i.e., changing the se-

Origins of cryptanalysis

quence of the characters of a text. When they evaluated a method they also considered how a potential attacker might analyse it. They were therefore aware of the significance of relative letter frequency in a language for the analysis of substitution ciphers, because it gave some insight into substitution rules. By the beginning of the fifteenth century, the Arabic encyclopaedia 'Subh al-a'sha' already contained an impressive treatment and analysis of cryptographic methods.

In Europe, cryptology originated during the Middle Ages in the papal and Italian city states. The first encryption algorithms merely involved vowel substitution, and therefore offered at least some rudimentary protection from totally ignorant attackers who may not have come up with the idea of trying out all the different possible vowel substitutions.

Protection of infrastructure

Not wanting to turn the entire development of cryptology into a scientific discipline at this juncture, we can deduce from the developments mentioned that special importance has always been given to protecting information. However, a second category of risks is increasingly becoming a major priority in the age of omnipresent communication networks. These risks actually affect communication infrastructures rather than the data being transmitted. With the development and expansion of increasingly complex networks, and the growing importance of these networks not only to the economic but also to the social development of the modern information society, there is also a greater demand for ways to secure communication infrastructures from deliberate manipulation. For economic operation it is important to ensure that the services provided by communication networks are available and functioning properly and that the use of these services can be billed correctly and in a way that everyone can understand.

1.1   Content and Structure of this Book

In this book equal treatment is given to the two task areas in network security mentioned – *security of transmitted data* and *security of the communication infrastructure*. We start by introducing central terms and concepts and providing an overview of the measures available for information security.

Part 1 of the book deals with fundamental principles

Building on this introductory information, the rest of the chapters in Part 1 deal with the *fundamental principles of data security technology*. Chapter 2 uses basic concepts to introduce cryptology. Chapter 3 covers the use and functioning of symmetric ciphering schemes, whereas Chapter 4 is devoted to asymmetric cryptographic algorithms. Chapter 5 introduces cryptographic check values for the detection of message manipulation. Generating secure, non-predictable random numbers is the subject of Chapter 6. In a sense, the algorithms in these four chapters constitute the *basic primitives* of data security technology upon which the cryptographic protection mechanisms of network security are based. Chapter 7 discusses cryptographic protocols and introduces the authentication and key exchange protocols that are central to network security. Part 1 concludes with Chapter 8, which provides an introduction to the principles of access control.

Part 2 of this book focuses on the architectures and protocols of *network security*. It starts with Chapter 9, which examines general issues relating to the integration of security services in communication architectures. Chapter 10 discusses security protocols of the data link layer, Chapter 11 examines the security architecture for the Internet protocol *IPSec*, and Chapter 12 describes the security protocols for the transport layer. Part 2 concludes with Chapter 13, which introduces *Internet firewalls* for realising subnetwork-related access control.

Part 2 introduces architectures and protocols for network security

The last part of the book, Part 3, presents the field of *secure wireless or mobile communication*. Chapter 14 differentiates the additional security aspects that arise in mobile communication compared with conventional fixed networks, and presents approaches of a more conceptual nature for maintaining the confidentiality of the current location area of mobile devices. The other chapters in this part examine concrete examples of systems. Chapter 15 deals with the security functions and weaknesses of the IEEE 802.11 standard for wireless local networks. Chapter 16 introduces the security functions for the two European standards for mobile wide-area networks *GSM* and *UMTS* and concluding Chapter 17 covers the fundamental aspects of secure mobile Internet communication based on the *Mobile IP* approach.

Part 3 is devoted to wireless and mobile communication

Before our attentive and inquisitive readers get too involved in the further content of this book, they should be made aware that the field of network security has developed into a very active field during the last few years. Consequently, extensive improvements are constantly being made to existing security protocols and new protocols are being developed and introduced. Doing justice to the speed of this development in a textbook therefore becomes a very difficult if not impossible undertaking. We therefore ask for the reader's understanding if a detail or two has already been

The field of network security is currently marked by a major dynamic

resolved in a way that deviates from our interpretation in a particular chapter or totally new protocols have established themselves in the meantime and are not dealt with in this book. It is precisely because of the rapid developments in this field that the priority of this book is to provide the reader with a fundamental understanding of the central principles presented and to describe them on the basis of concrete and relevant sample protocols.

1.2 Threats and Security Goals

The terms *threat* and *security goal* play an important role in assessing the risks in communication networks. Therefore, they will first be defined in general terms.

Definition 1.1 *A* **threat** *in a communication network is a potential event or series of events that could result in the violation of one or more security goals. The actual implementation of a threat is called an* **attack***.*

Examples of concrete threats

Definition 1.1 given above is kept quite abstract and refers to the term security goal defined below. The following examples clarify the types of threats that exist:

- ❏ a 'hacker' intruding into the computer of a company;

- ❏ someone reading someone else's transmitted e-mails;

- ❏ a person altering sensitive data in a financial accounting system;

- ❏ a hacker temporarily shutting down a web site;

- ❏ somebody using or ordering services and goods in someone else's name.

Examples of security goals

The term *security goal* is another concept that is easier to explain with examples because at first glance security goals can vary considerably depending on the respective application scenario:

- ❏ Banks:

    - ❏ protection from deliberate or unintentional modification of transactions;

    - ❏ reliable and non-manipulable identification of customers;

❏ protection of personal identification numbers from disclosure;

❏ protection of personal customer information.

❏ Administration:

❏ protection from disclosure of sensitive information;

❏ use of electronic signatures for administrative documents.

❏ Public network operators:

❏ restriction of access to network management functions to authorised personnel only;

❏ protection of the availability of the services offered;

❏ guarantee of accurate and manipulation-safe billing of use of services;

❏ protection of personal customer data.

❏ Corporate and private networks:

❏ protection of the confidentiality of exchanged data;

❏ assurance of the authenticity of messages (see below).

❏ All networks: Protection from intrusion from outside.

Some of the security goals listed above are of course relevant to several different application scenarios (even if they are not repeated in the categories above). However, security goals can also be defined from a purely technical standpoint without being based on a concrete application scenario.

General definition of security goals

Definition 1.2 *A distinction can generally be made between the following* **technical security goals***:*

❏ **Confidentiality:** *Transmitted or stored data should only be disclosed to authorised entities.*

❏ **Data integrity:** *It should be possible to detect unintentional or deliberate changes to data. This requires that the identification of the originator of the data is unique and cannot be manipulated.*

❏ **Accountability:** *It must be possible to identify the entity responsible for a particular event (e.g., use of a service).*

❑ **Availability:** *The services implemented in a system should be available and function properly.*

❑ **Controlled access:** *Only authorised entities should be able to access certain services and data.*

General technical threats
Like security goals, threats can be viewed from a primarily technical standpoint and therefore *technical threats* are distinguished as follows:

❑ *Masquerade:* An entity pretends to have the identity of another entity.

❑ *Eavesdropping:* An entity reads information that is meant for someone else.

❑ *Authorisation violation:* An entity uses services or resources although it does not have appropriate permission.

❑ *Loss or modification of information:* Certain information is destroyed or modified.

❑ *Forgery:* An entity creates new information using the identity of another entity.

❑ *Repudiation:* An entity falsely denies having participated in a particular action.

❑ *Sabotage:* Any action that is aimed at reducing the availability or correct functioning of services or systems.

These terms can be used as the basis for creating a general classification that clarifies which security goals are in danger of being exposed to which threats. Table 1.1 provides an overview of this classification. The table can be read in two different ways. On one hand, it shows that information confidentiality is threatened by the technical threats of masquerade, eavesdropping and authorisation violation; on the other hand, it can also be directly inferred from the table that forgery primarily threatens the security goals of data integrity, accountability and controlled access.

Real attacks often combine several threats
In reality, a concrete attack often involves a combination of the threats mentioned above. An intrusion into a system often involves sniffing the access identification and related password. The identity of the sniffed identification is then provided for the access check with the latter representing a masquerade.

Technical Security Goals	Technical Threats						
	Masquerade	Eavesdropping	Authorisation Violation	Loss or Modification of Information	Forgery of Information	Repudiation of Events	Sabotage (e.g. by Overload)
Confidentiality	x	x	x				
Data Integrity	x		x	x	x		
Accountability	x		x	x		x	
Availability	x		x	x			x
Controlled Access	x		x		x		

Table 1.1
Technical Security Goals and Threats

1.3   Network Security Analysis

When appropriate action is taken to counteract the above-mentioned threats to an actual application scenario, the counter-measures being considered first have to be evaluated carefully for the given network configuration. This requires a detailed *security analysis* of the network technology with an assessment of the risk potential of technical threats to the entities communicating in the network, along with an evaluation of the cost – in terms of resources and time (computing capacity, storage, message transfer) – of executing known attack techniques.

Note: Unknown attack techniques are generally not possible to evaluate!

Sometimes the detailed security analysis of a given network configuration or a specific protocol architecture can be used to convince an organisation's financial controlling of the need for essential security measures.

As a rule, a detailed security analysis of a specific protocol architecture can be structured according to the following finely granulated *attacks at the message level*:

❑ Passive attacks: Eavesdropping on protocol data units (PDUs);

❑ Active attacks: Delay, replay, deletion and insertion of PDUs.

One basic assumption in this context is that an actual hacker would have to be able to combine the attacks listed above in order to use them to construct more complex attacks from these ba-

Combination of attacks

sic building blocks interpreted as attack primitives. A 'successful attack' at the message level therefore requires that:

❏ The attack produces no detectable side effects for other communication processes, e.g., for other connections or connectionless data transmission.

❏ The attack produces no side effects for other PDUs in the same connection or in connectionless data transmission between the entities participating in the communication.

When a security analysis is produced for a protocol architecture, each individual layer in the protocol architecture should be checked for the attacks mentioned above.

Figure 1.1 shows the layered architecture typically used in communication systems today. In this architecture the end systems communicate with one another over a network of intermediate systems. The protocol functions are organised into five layers:

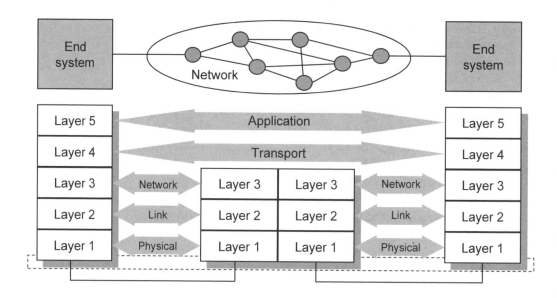

Figure 1.1
Architecture of layered communication systems

❏ The lowest layer is the *physical layer*, which is responsible for transmitting bit streams over a physical medium (e.g. line, radio transmission link).

❏ The *data link layer* above it combines multiple bits from the transmitted bit stream into transmission frames and carries out transmission that is secure from error between two systems connected over a physical medium. It performs two basic tasks: When a shared medium is available to several systems, it coordinates access to the shared medium *(Medium Access Control, MAC)*. It also takes appropriate measures to detect transmission errors so that defective frames received at the receiver are detected and can be discarded.

❑ The *network layer* is responsible for the communication between end systems that are normally linked to one another over several intermediate systems. Therefore, the main task of this layer is routing through the transmission network between the two end systems.

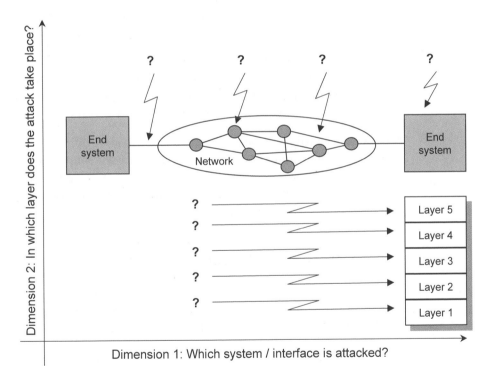

❑ The *transport layer* enables an exchange of data between the processes of the end systems. The key tasks of this layer are addressing applications processes, detecting errors at the end-to-end level and, with a reliable service, implementing measures for error recovery (e.g. through retransmission).

Figure 1.2
Dimensions of the security analysis of layered protocol architectures

❏ Above the transport layer the *application layer* – as its name suggests – implements applications-specific protocols that are as diverse as the applications run in the end systems.

Only the three lower layers up to the network layer are normally implemented in the (intermediate) systems of the transmission network.

Security analysis   According to the description given above, a security analysis of *structure*   layered protocol architectures can be structured along two dimensions (also compare Figure 1.2):

❏ First the *systems and interfaces at risk* in the network configuration being analysed must be identified. For example, publicly accessible end systems, gateways to public networks as well as non-secure transmission routes (particularly in the case of wireless transmission) pose special security risks.

❏ The security analysis is also structured according to the *layer* in which an attack can take place. Attacks do not necessarily have to occur in the application layer. On the contrary, depending on the intentions of the hacker, the main attack point can be the layers below the transport layer.

A detailed security analysis is very useful for identifying the security risks that dominate in a particular network configuration. It can be used as the basis for selecting appropriate security measures to reduce these risks. The following section provides a general overview on this subject.

1.4 Information Security Measures

Many different security measures are available, each dealing with specific aspects of an information processing system and its embedding into the work processes supported by the system:

❏ *Physical security measures* include lock systems and physical access controls, tamper proofing of security-sensitive equipment and environmental controls such as motion detectors, etc.

❏ *Personnel security measures* begin with a classification of the security-specific sensitivity of a position and also include procedures for employee screening and security training and awareness.

❑ *Administrative security measures* include procedures for the controlled import of new software and hardware, detection of security-relevant occurrences through maintenance and regular checks of event logs as well as an analysis of known security breaches and incidents.

❑ *Media security measures* are aimed at safeguarding the storage of information. Procedures and control mechanisms are implemented to identify, reproduce or destroy sensitive information and data carriers.

❑ *Radiation security measures* designed to prevent or limit electromagnetic emission from computer systems and peripheral devices (especially monitors) that a hacker could note and use to eavesdrop on information.

❑ *Life-cycle controls* monitor the design, implementation and introduction of information processing systems. The specification and control of standards to be upheld for programming and documentation are geared towards achieving a 'reliable' development process.

❑ *System security measures* for computers, operating systems and the applications run on computers are designed to secure information that is stored and processed in computing systems.

❑ Expanding on the latter category, *communication security measures* are designed to protect information while it is being transmitted in a communication network. In conjunction with the measures that protect the network infrastructure itself, they form the category of *network security measures*.

The last category mentioned, network security, is the main subject of this book. However, it should be emphasised that a careful application of the entire catalogue of measures listed above is necessary to guarantee the security of information processing processes. This is due to the fact that a security system is only as secure as its weakest component. For example, a sophisticated password system that prevents the use of easily guessed passwords is minimally effective if users write their passwords on media that are not adequately protected or if a hacker can use a telephone call to induce someone to divulge a password ('social engineering').

A secure information processing process requires a comprehensive catalogue of measures

1.5   Important Terms Relating to Communication Security

This section introduces the terms *security service, cryptographic algorithm* and *cryptographic protocol*, which are central to network security, and explains their relationship to one another.

Definition 1.3 *A* **security service** *is an abstract service that seeks to achieve a specific security objective.*

Implementation of security services

A security service can be implemented through either cryptographic or conventional means. For example, one way to prevent a file stored on a disk from being read by an unauthorised entity is by ensuring that the file is encrypted before it is stored. On the other hand, the same goal can be achieved if the disk is locked up in a secure safe. Normally, the most effective approach is a combination of cryptographic and conventional methods.

Fundamental security services

In its generalisation, Definition 1.3 gives the impression that a multitude of different security services exist. Actually the number is surprisingly small; precisely five fundamental security services are distinguished:

❏ As subsequent discussions in this book will show, *authentication* is the most important of all security services, because it allows manipulation-safe identification of entities.

❏ To a certain extent the security service *data integrity*, which ensures that data generated by a specific entity cannot undetectably be modified, is 'the small brother' of the authentication service.

❏ *Confidentiality*, which is aimed at preventing information from being made known to unauthorised entities, is probably the most popular security service.

❏ The security service *access control* checks that every entity that has proper authorisation can access certain information and services in a specified way.

❏ The aim of the *non-repudiation* service is to enable a unique identification of the initiators of certain actions, such as the sending of a message, so that these completed actions cannot be disputed after the fact.

Definition 1.4 *A* **cryptographic algorithm** *is a mathematical transformation of input data (e.g. data, keys) to output data.*

Cryptographic algorithms play an important role in the realisation of security services. However, a cryptographic algorithm used on its own is not sufficient because it also has to be embedded in a semantic context. This usually occurs as part of the definition of a *cryptographic protocol*.

Definition 1.5 *A* **cryptographic protocol** *is a procedural instruction for a series of processing steps and message exchanges between multiple entities. The aim is to achieve a specific security objective.*

The last two terms defined for cryptographic algorithms and protocols are of such fundamental significance for network security that they are dealt with in several chapters. However, the next chapter will first introduce the general basics of cryptology.

2 Fundamentals of Cryptology

This chapter introduces the basic concepts of cryptology [Sim94a, Riv90]. The first section starts with a definition of the general terms cryptology, cryptography and cryptanalysis. Section 2.2 follows with a basic classification of the cryptographic algorithms that occur within the area of network security. Sections 2.3 and 2.4 introduce basic issues relating to cryptanalysis and Section 2.5 examines the properties of encryption schemes important to communication and presents a breakdown of the different schemes. Lastly, Section 2.6 looks at the main tasks involved in the management of cryptographic keys, which is of central importance in the use of cryptographic methods.

2.1 Cryptology, Cryptography and Cryptanalysis

The term *cryptology* refers to the science of secure and, as a rule, confidential communication. The word itself derives from the two Greek words *kryptós* (hidden) and *lógos* (word). Cryptology comprises two main areas:

❏ *Cryptography* analyses and develops methods for transforming unsecured *plaintext* into *ciphertext* that cannot be read by unauthorised entities. The term is made up from the two words *kryptós* and *gráphein* (writing). As the field has developed over the last twenty or so years, additional categories of algorithms have been added to the main subject of pure encryption schemes. These algorithms are generally also considered cryptographic algorithms although they do not realise encryption in a true sense.

❏ In some respects, *cryptanalysis*, sometimes also called *cryptological analysis*, represents the antagonist of cryptography, because it is the science' and partly also the art' of recovering plaintext from ciphertext without knowledge of the key

Cryptanalysis is a complementary discipline to cryptography

that was used. The term comes from the two Greek words *kryptós* and *analýein* (loosening, untying). Cryptanalysis is an essential extension of cryptography, which means that a cryptographer has always also to be a cryptanalyst so that the methods he or she develops can actually accomplish what is expected of them.

2.2   Classification of Cryptographic Algorithms

Main applications of cryptography

There are two main applications of cryptography that are important in connection with network security:

❏ Data *encryption*, which is a transformation of plaintext into ciphertext, as well as the inverse operation *decryption*, sometimes also referred to as *ciphering* and *deciphering*, respectively.

❏ Data *signing*, which is a manipulation-safe calculation of the checksum. On the basis of a *signature check*, the checksum makes it possible to determine whether data was modified, either due to errors or deliberate manipulation, after it was created and signed. It is important to make a distinction between *signing* as it is defined here and the generating of a *digital signature*.

Some cryptographic algorithms can be used for both applications, whereas others are secure and/or can be efficiently used for only one of the applications.

Classification according to number of keys

At a general level, cryptographic algorithms can be classified according to how many different keys they use:

❏ *Symmetric cryptographic algorithms* use *one* key, which means that a message is decrypted with the same key that was used to encrypt it or a signature is checked with the same key that created it.

❏ *Asymmetric cryptographic algorithms* use *two* different keys for encryption and decryption or signing and signature check. The two keys cannot be selected independently of the other and instead must be constructed as a pair specifically in accordance with the cryptographic algorithm.

❏ *Cryptographic hash functions* use *no* key. This means that they implement a transformation from input data that is defined only by the specific algorithm but not through an additional key. This category of algorithms is mainly used to create cryptographic check values (see Chapter 5) and to generate pseudo-random numbers (see Chapter 6).

A separate chapter is devoted to each of these categories. However, we will first present some general observations on cryptanalysis and on the characteristics and classification of encryption algorithms.

2.3 Cryptanalysis

Cryptanalysis is essentially the process used to obtain the corresponding plaintext or appropriate key from ciphertext. Depending on the cryptanalyst's requirements, a distinction is made between the following classic types of cryptanalysis:

❏ *Exhaustive search of key space*: Also called *brute-force attack*, this is basically the simplest but usually also the most time consuming form of attack. First it makes an assumption about the algorithm used to encrypt a plaintext and then successively tries out all possible keys to decrypt an intercepted ciphertext. If the ciphertext can be represented in 'intelligible' plaintext with one of the keys, it is then assumed that the key being sought has been discovered. Statistically this involves searching through half the key space. Because this form of attack can basically be used against every cryptographic scheme, it imposes an upper limit for the effort required to break arbitrary cryptographic schemes. The effort required increases exponentially with the length of the key.

Brute-force attack

❏ *Analysis based only on known-ciphertext*: This form of cryptanalysis is based on the assumption that certain properties of a plaintext will be retained even with encryption. For example, simple ciphers retain the relative frequency of individual letters or combinations of letters. With this attack form, different ciphertexts $C_1, C_2, ...$, all created with the same key K, are available to the cryptanalyst. This means that plaintexts $P_1, P_2, ...$ exist so it holds that $C_1 = E(K, P_1), C_2 = E(K, P_2), ...$ with $E(K, P)$ representing the encryption of plaintext P with key K. Thus the objective of

Known-ciphertext

the cryptanalyst is to determine either as many plaintexts $P_1, P_2, \ldots$ as possible or even the key K.

Known-plaintext ❑ *Analysis based on known-plaintext pairs*: With this form of analysis the cryptanalyst has access to a set of known-plaintext pairs $(P_1, C_1), \ldots, (P_i, C_i)$, all of which have been generated with the same key K. The objective is to discover key K or at least an algorithm that can be used to obtain the plaintexts of other ciphertexts $C_{i+1}, \ldots$.

Chosen plaintext ❑ *Analysis based on chosen plaintext*: In this case, the cryptanalyst not only has access to pairs of plaintexts and ciphertexts generated with the same key, but he can also choose plaintexts $P_1, P_2, \ldots$ himself and recover the corresponding ciphertexts $C_1, C_2, \ldots$ that were encrypted with the same key K. The objective of this analysis is to determine the unknown key K or at least an algorithm for decrypting other ciphertexts from the pairs (P_i, C_i).

Chosen ciphertext ❑ *Analysis based on chosen ciphertext*: This attack form is very similar to the previous one. The only difference is that the cryptanalyst specifies ciphertexts $C_1, C_2, \ldots$ and recovers the corresponding plaintexts $P_1, P_2, \ldots$ that are decrypted with the same key K. Here again the objective of the analysis is to determine the unknown key K or at least an algorithm for decrypting other ciphertexts from the pairs (P_i, C_i).

Unpublished cryptographic methods usually do not offer sufficient security An important assumption with all the forms of analysis listed is that the cryptanalyst knows how the cryptographic algorithm functions. This assumption, which is now generally accepted as being a sensible one, relates to the principle named after the Flemish cryptographer August Kerckhoff: It states that the strength of a cryptographic method must be based completely on the secrecy of the key and not the secrecy of the algorithm [Ker83]. This principle is supported by the fact that extensive cryptanalytic studies by a large number of experienced cryptanalysts provides a statement about the security of a cryptographic algorithm. A popular aphorism among cryptographers states that 'anyone can design a cryptographic algorithm that he cannot break himself.'

New developments in the field have led to the techniques of *differential cryptanalysis* [BS90, BS93] and *linear cryptanalysis* [Mat94]. Both techniques are specialisations of the basic forms discussed above: differential cryptanalysis implements a special form of chosen plaintext analysis and linear cryptanalysis is a special form of known-plaintext analysis.

The cryptanalysis of asymmetric cryptographic methods can- *Cryptanalysis of*
not be organised into the classification just described, because it *asymmetric methods*
usually starts at a special place: the construction of matching key
pairs. As the section on asymmetric cryptography explains in detail, these key pairs are constructed using mathematical attributes
specified by the concrete cryptographic algorithm. The cryptanalyst tries to exploit these attributes by computing the appropriate
private key from the public key of the two keys. Cryptanalysis of
asymmetric methods therefore actually represents a separate discipline of mathematical research – more accurately, a subfield of
number theory, and works with means different from those in the *Asymmetric*
classic cryptanalysis of symmetric methods. Overall it is directed *cryptanalysis is aimed*
more towards completely breaking cryptographic algorithms than *at completely*
only calculating some plaintexts or keys from actual ciphertexts. *breaking algorithms*

2.4 Estimating the Effort Needed for Cryptographic Analyses

As mentioned above, the easiest way to attack any cryptographic *Which circumstances*
algorithm is by systematically trying out all potential keys. For *and which means are*
example, all possible keys are used systematically to decrypt a ci- *needed to carry out a*
phertext until an 'intelligible' plaintext is found. The probability *specific cryptanalysis?*
is relatively high that the key discovered this way is the right key,
and the cryptanalyst will be fairly certain of having found the right
key by the time a second ciphertext is encrypted into intelligible
plaintext with the same key. Since the number of possible keys
with all the methods being used today is finite, this approach will
inevitably be successful *sometime*. This section will look into the
question of 'when' this sometime will be and related issues.

An estimation of the effort required for an attack on a cryptographic method usually includes the following:

❑ An estimate of the average number of 'steps' needed, with one
   step possibly consisting of a sequence of complex calculations.

❑ An assumption of the number of steps that can be executed in
   one time unit and usually also based on a technical assumption, e.g. a computation by semiconductor elements possibly
   calculated on the high side with generally recognised estimation such as Amdahl's Law, which basically estimates that
   the achievable processing speed doubles every 18 months.

❑ An estimate of the memory needed for an analysis.

Goal is an estimate of
the dimension of the
effort
The estimates needed are not an exact calculation of the actual
effort. What is of interest instead is an assessment of the magnitude involved. Additive and multiplicative constants are therefore
not important. What is important is only the term that calculates
the magnitude, and with algorithms for cryptographic attacks it is
usually exponentially in the length of the key used.

For example, for the brute-force attack described earlier, one
work step involves identifying the key being checked in this step,
decrypting the recorded ciphertext, and checking whether the
plaintext recovered from it is intelligible. To simplify matters
more, it is usually assumed that only the decryption is the dominant value.

Table 2.1
*Average times needed
for an exhaustive
search for a key*

Key length [bit]	Number of keys	Time req. with 1 encrypt. / μs	Time req. with 10^6 encrypt. / μs
32	$4.3 \cdot 10^9$	35.8 min.	2.15 ms
56	$7.2 \cdot 10^{16}$	1142 years	10.01 years
128	$3.4 \cdot 10^{38}$	$5.4 \cdot 10^{24}$ years	$5.4 \cdot 10^{18}$ years

Table 2.1 lists the resulting times of a brute-force attack for different key lengths and different decryption speeds. Note that the
value of 10^6 decryptions per microsecond (μs) was realistic for the
DES algorithm and the technology available in the mid-1990s.
*Estimating the
development of
processing times
using Amdahl's Law*
Amdahl's Law can be used for a relatively reliable estimate of the
development of these processing times because no changes result
to the main switching logic required for an attack. What should
be exempted explicitly from such estimates are principal scientific
developments in respect to the underlying computation model. A
principal example is the methods of *quantum informatics* that are
not yet ready for practical implementation but are theoretically
designed [CU98].

*The practicality of
cryptanalytic schemes
can be evaluated
based on a
comparison of time
and material
requirements against
reference values of
the world around us*
When such results are evaluated, it is useful to know some
of the comparison values that relate the computed value to the
magnitudes of the world around us. Table 2.2 lists some of these
values.

What the two tables show is that the time needed for a brute-force calculation of a 128-bit long key exceeds the age of our solar
system. The assumption that such an attack is not practical is
therefore justified.

The estimated number of electrons that exist in our universe
is another useful comparison value, because it represents – even if
the number is on the high side – an upper limit for the maximum

	Reference	Magnitude
	Seconds in a year	$3 \cdot 10^7$
	Seconds since creation of solar system	$2 \cdot 10^{17}$
	Clock cycles per year at 50 MHz	$1.6 \cdot 10^{15}$
	Binary numbers of length 64	$1.8 \cdot 10^{19}$
	Binary numbers of length 128	$3.4 \cdot 10^{38}$
	Binary numbers of length 256	$1.2 \cdot 10^{77}$
	Number of 75-digit prime numbers	$5.2 \cdot 10^{72}$
	Number of electrons in the universe	$8.4 \cdot 10^{77}$

Table 2.2
Reference values for estimating the computational effort of cryptanalytic methods

memory space available to an attacker. This is of course based on a technological assumption that at least one electron is needed to store an information unit.

The table contains one last important value, the magnitude of the number of 75-digit prime numbers. The significance of prime numbers to asymmetric cryptography is clarified in Chapter 4. The only important point to mention here is that prime numbers exist in sufficient number for the probability that the same prime number will be selected twice during a random selection of large prime numbers to be minimal.

2.5 Characteristics and Classification of Encryption Algorithms

This section introduces two characteristics of encryption algorithms that are important for message transfer, and presents an initial categorisation of these schemes.

Assume that a sender wants to send a series of plaintexts $P_1, P_2, \ldots$ to a receiver. Because the messages are confidential, the sender first encrypts them into ciphertexts $C_1, C_2, \ldots$ and then transmits these to the receiver. There are two characteristics of the encryption method that are important:

- ❏ *Error propagation* in an encryption scheme characterises the effect of bit errors during the transmission of ciphertext on plaintexts $P_1', P_2', \ldots$ reconstructed after decryption. Depending on the cryptographic method used, a bit error in the ciphertext can result in a bit error or even a large number of defective bits in the plaintext.

Error propagation

The error propagation of a cryptographic method should be considered particularly in connection with *error-correcting codes*. By transmitting redundant information such codes enable certain errors, e.g. single-bit errors or even bursty errors up to a certain length, to be corrected automatically at the receiver. When they are used simultaneously with an encryption method that propagates transmission errors, it is important that encryption takes place first before the redundant information of the error-correcting code is computed. This ensures the error-correcting properties of the code, thereby also reducing the probability of errors in defective recovered plaintexts.

Synchronisation feature

❏ In contrast, the *synchronisation feature* of an encryption scheme describes the effect of a lost ciphertext C_i on plaintexts $P_{i+1}, P_{i+2}, ...$ reconstructed from subsequent correctly transmitted ciphertexts $C_{i+1}, C_{i+2}, ...$ Depending on the method used, it can turn out that such a loss has no effect on following plaintexts, that some subsequent plaintexts are falsified or even that all the following ciphertexts are incorrectly deciphered, and therefore explicit synchronisation is required between sender and receiver.

At the topmost level, encryption schemes can be categorised according to the following dimensions:

Type of operations

❏ According to the type of operations used for mapping plaintext to ciphertext:

❏ *Substitution* maps elements of the plaintext, e.g. bits, letters or groups of bits or letters, to other elements.

❏ *Transposition* rearranges the sequence of the elements of a plaintext.

Most encryption methods today combine both basic techniques because pure substitution or transposition ciphers do not offer adequate security.

Number of keys used

❏ According to the number of keys used:

❏ *Symmetric encryption methods* use the same key for decryption as for encryption.

❏ *Asymmetric encryption methods* use two different but matching keys for the two complementary operations.

❏ According to the way in which the plaintext is processed: *Type of processing*

❏ *Stream ciphers* work on bit streams, which means that they encrypt plaintext one bit after another. Many stream ciphers are based on the idea of linear feedback shift registers that realise very efficient encryption. However, a multitude of cryptographic weaknesses has already been discovered in this method because of the sophisticated mathematical theory that exists for this field. The use of stream ciphers based on feedback shift registers is therefore often discouraged [Sch96, Chapters 16 and 17]. Most stream ciphers do not incorporate error propagation but instead react with great sensitivity to any loss of synchronisation and in such cases require explicit resynchronisation.

❏ *Block ciphers* transform blocks of length b bits with the parameter b always dependent on the actual algorithm used.

2.6 Key Management

If cryptographic measures are used to protect data in a system, the management of the keys required for this purpose becomes a critical task. This is because data loses its cryptographic protection if the keys used to encrypt it are known. The tasks involved in key management are discussed in detail below:

❏ *Key generation* is the creation of the keys that are used. This *Key generation* process must be executed in a *random* or at least *pseudo-random-controlled* way, because hackers will otherwise be able to execute the process themselves and in a relatively short time, will discover the key that was used for security. Pseudo-random-controlled key generation means that keys are created according to a deterministic approach but each possible key has the same probability of being created from the method. Pseudo-random generators must be initialised with a real random value so that they do not always produce the same keys. If the process of key generation is not reproducable, it is referred to as 'really random' key generation. The generation of random numbers is dealt with at length in Chapter 6.

If keys are needed for a symmetric cryptographic method, the output of a random or pseudo-random generator can be deployed as a key. On the other hand, key pairs produced for asymmetric cryptographic algorithms based on mathematical problems of factorisation or discrete logarithms require the generation of random and large prime numbers [Bre89].

Key distribution ❏ The task of *key distribution* consists of deploying generated keys in the place in a system where they are needed. In simple scenarios the keys can be distributed through direct (e.g. personal) contact.

If larger distances are involved and symmetric encryption algorithms are used, the communication channel again has to be protected through encryption. Therefore, a key is needed for distributing keys. This necessity supports the introduction of what is called *key hierarchies* . The three-tier hierarchy of the IBM key management scheme [DP89, pages 143 ff.] is an example. The scheme distinguishes between two *master keys* at the top level, a number of *terminal keys* at the next level and, lastly, *session keys* at the third level. Public keys of asymmetric cryptographic schemes (compare Chapter 4) generally can be made known. They only require that authenticity be secured but not the confidentiality.

Key storage ❏ When *keys are stored*, measures are needed to make sure that they cannot be read by unauthorised users. One way to address this requirement is to ensure that the key is regenerated from an easy to remember but sufficiently long password (usually an entire sentence) before each use, and therefore is only stored in the memory of the respective user. Another possibility for storage is manipulation-safe crypto-modules that are available on the market in the form of processor chip cards at a reasonable price [RE95, VV96].

Key recovery ❏ *Key recovery* is the reconstruction of keys that have been lost. The simplest approach is to keep a copy of all keys in a secure place. However, this creates a possible security problem because an absolute guarantee is needed that the copies of the keys will not be tampered with [Sch96, pages 181 f.]. The alternative is to distribute the storage of the copies to different locations, which minimises the risk of fraudulent use so long as there is an assurance that all parts of the copies are required to reconstruct the keys.

❏ *Key invalidation* is an important task of key management, particularly with asymmetric cryptographic methods. If a private key is known, then the corresponding public key needs to be identified as invalid. This task is complicated by the fact that public keys are possibly stored in public directories worldwide and numerous but non-accessible copies of these keys may exist.

Key invalidation

❏ The *destruction of no longer required keys* is aimed at ensuring that messages ciphered with them also cannot be decrypted by unauthorised persons in the future. It is important to make sure that all copies of the keys have really been destroyed. In modern operating systems this is not a trivial task since storage content is regularly transferred to hard disk through automatic storage management and the deletion in memory gives no assurance that copies of the keys no longer exist. In the case of magnetic disk storage devices and so-called EEPROMs (Electrically Erasable Programmable Read-Only Memory), these have to be overwritten or destroyed more than once to guarantee that the keys stored on them can no longer be read, even with sophisticated technical schemes.

Destruction of no longer required keys

2.7 Summary

Cryptology is the science of secure communication based on an algorithmic transformation of communicated data. It consists of the two fields of *cryptography* and *cryptanalysis*. Two main applications are of special interest in the context of network security: data *encryption* to hide its meaning and data *signing*, which is the generation of cryptographically secure check values to check the authenticity of a message.

Main cryptography applications for network security are encryption and signing

A number of different cryptanalysis methods exist and they are usually categorised according to which information is available to cryptanalysts for an attack. The simplest form of attack consists of trying out all possible keys. Since the effort required for this form of attack grows exponentially according to key length, a hacker will reach the physical limits of our universe if the keys are long enough.

The following chapters examine a range of cryptographic algorithms that are of central importance to network security. We have purposely limited the selection because it is not the goal of this

This book presents an essential selection of algorithms

Figure 2.1
*Overview of
cryptographic
algorithms presented
in this book*

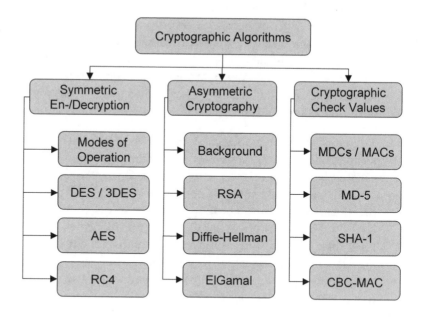

book to present a comprehensive introduction to the field of cryptology. Our aim instead is to provide a fundamental understanding of how cryptographic methods work as the basis for explaining their use in protocols and architectures of network security.

*Content of the
chapters on
cryptographic
algorithms*

Figure 2.1 presents an overview of the algorithms discussed in this book. The following chapter deals with symmetric encryption methods and Chapter 4 is devoted to asymmetric cryptographic algorithms. Chapter 5 provides an introduction to the field of cryptographic checksums.

2.8   Supplemental Reading

[Ker83]   KERCKHOFF, A.: La Cryptographie Militaire. In: *Journal des Sciences Militaires* (1883), January 1883
The important principle stating that a cryptographic algorithm has to provide security even if its functioning has been made known to a possible adversary was first formulated in this historically interesting article on the state of technology in the field of encryption in the late Nineteenth century.

[Riv90]   RIVEST, R.: Cryptography. In: VAN LEEUWEN, J. (Ed.):
           Handbook of Theoretical Computer Science Vol. 1, Else-
           vier, 1990, pp. 717–755

[Sim94a]   SIMMONS, G. J.: Cryptology. In: *Encyclopaedia Britan-
           nica*, Britannica, 1994

2.9 Questions

1. How much computation is required for a brute-force attack on
    a ciphertext in an optimal case, in an average case and in the
    worst case?

2. Can it make sense for error-correcting coding to be provided
    in layer $n + 1$ of a protocol tower and encryption in layer n? If
    yes, which requirements have to be met?

3. The *Caesar-cipher* named after Julius Caesar shifts every let-
    ter in the alphabet upward by three positions (thus 'A' be-
    comes 'D,' 'B' becomes 'E', etc.). How many possible keys can
    be used? How many keys are theoretically possible with this
    scheme?

4. Can there be any benefit to a cryptographic algorithm that is
    designed with a finite length key but is based on a theoreti-
    cally undecidable problem of computer science?

3 Symmetric Cryptography

Symmetric encryption schemes use the same key for enciphering and deciphering. This key must be kept secret and should only be known to entities that need the key to read encoded ciphertexts. This chapter explains the basic encryption modes of symmetric encryption algorithms and describes DES, AES and RC4, currently the most-widely used algorithms.

3.1 Encryption Modes of Block Ciphers

In Section 2.5 we established that block ciphers do not encrypt or decrypt a bit stream bit by bit but instead in b length blocks, with the block length b specified by the encryption algorithm. This leads us to the question of how messages with a length different than b can be encrypted. To keep the discussion simple, we will assume below that the length of the message is a multiple of the block length b. This assumption can be applied to any messages that use known data processing methods, e.g., adding a '1' bit and a variable number of '0' bits until reaching the next multiple of b.

How can messages with a different length than the block size of the cipher be encrypted?

At this point one could come to the conclusion that message *padding* would completely solve the problem of the encryption of different length messages: all that is still needed is that each block of the length b has to be encrypted individually. Actually this describes one possible procedure. However, the approach has a serious cryptographic drawback and, as a result, alternative procedures were developed. These procedures are generally referred to as *encryption modes* and are explained below. First the following notation, which is also valid in the following chapters, is agreed.

Using padding alone to fill blocks and block-wise message processing is not enough

Because keys for symmetric schemes have to be kept secret, they are normally always agreed between two entities A and B. When more than one key is involved, it is useful if the two entities are noted as the index of the key so that a key can be identified as $K_{A,B}$, for example. However, this indexing is often dispensed with when it is implicitly clear or immaterial which entities have

Notation

agreed key K. The encryption of a plaintext P into a ciphertext C is often noted as $C = E(K_{A,B}, P)$ through the function E for encrypt. Its complementary operation is expressed by the function D for decrypt and it holds that: $D(K_{A,B}, E(K_{A,B}, P)) = P$. For space reasons the notations $E_{K_{A,B}}(P)$ or, even shorter, $\{P\}_{K_{A,B}}$ are sometimes used to denote encryption of a message.

A plaintext message P is segmented into blocks $P_1, P_2, \dots$ of the same length j, with $j \leq b$ and j normally a factor of b. The corresponding ciphertext blocks are of the same length and are noted with $C_1, C_2, \dots$.

Electronic Code Book Mode

The procedure described above in which a message P is encrypted through a separate enciphering of the individual message blocks P_i is called *Electronic Code Book Mode (ECB)* and is illustrated in Figure 3.1.

If an error occurs during the transmission of a ciphertext block C_i, the entire decrypted plaintext block P_i' is always falsified. However, the blocks that follow are correctly decrypted again and no subsequent errors occur even if one or more ciphertext blocks are lost. On the other hand, if the number of 'lost' bits has a value other than the block size b of the encryption procedure, then explicit resynchronisation is required between sender and receiver. The main disadvantage of this scheme is that identical plaintext blocks are mapped to identical ciphertext blocks. This characteristic makes the procedure vulnerable to a number of cryptographic analysis schemes and generally, it therefore should not be used.

Figure 3.1
Electronic Code Book Mode

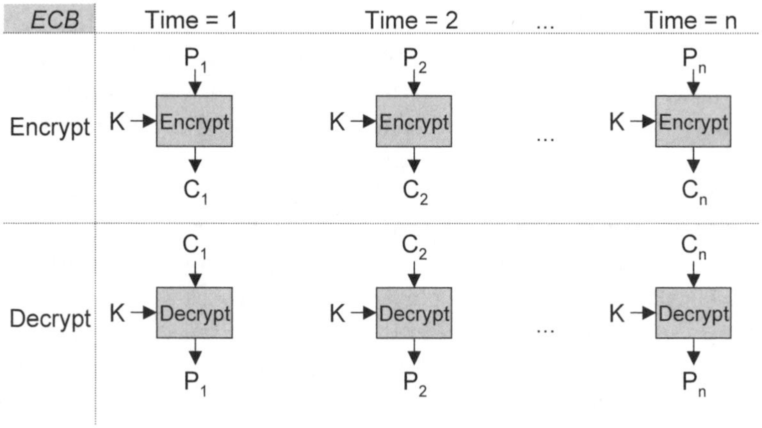

Cipher Block Chaining Mode

Mapping the same plaintext blocks to identical ciphertexts should be avoided. Therefore, with *Cipher Block Chaining Mode (CBC)* plaintext blocks P_i are XORed to previous ciphertext block C_{i-1} before encryption. Consequently, when the blocks are decrypted, the value obtained after decryption of received ciphertext C_i still has to be XORed with the previous ciphertext C_{i-1} to recover the correct plaintext P_i. An *initialisation vector (IV)* is agreed as value C_0 between sender and receiver for the encryption and decryption of the first message block. This value does not have to be hidden from attackers and can be transmitted in plaintext before the actual message is sent. An overview of this mode is shown in Figure 3.2 and the following equations give additional clarification of the relationships described:

$$
\begin{aligned}
C_i &= E_K(C_{i-1} \oplus P_i) \\
D_K(C_i) &= D_K(E_K(C_{i-1} \oplus P_i)) \\
D_K(C_i) &= C_{i-1} \oplus P_i \\
C_{i-1} \oplus D_K(C_i) &= C_{i-1} \oplus C_{i-1} \oplus P_i \\
C_{i-1} \oplus D_K(C_i) &= P_i
\end{aligned}
$$

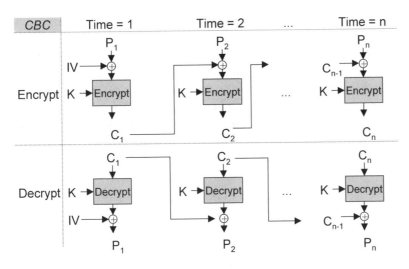

Figure 3.2
Cipher Block Chaining Mode

If one or more bit errors occur in ciphertext block C_i, the entire deciphered plaintext block P_i' is falsified, as are the corresponding bits in following block P_{i+1}'. The loss of ciphertext block C_i also results in a falsification of the following block P_{i+1}'. However, the scheme resynchronises itself automatically after one block has been incorrectly deciphered as long as complete blocks are lost.

In some instances message blocks even shorter than those with a b bit length require encryption. An example is confidential

Processing shorter message blocks

interactive terminal communication where individual characters should ideally be encrypted and then transmitted immediately after being typed in by the user.

Using padding alone
to achieve block size
wastes bandwidth and
leads to cryptographic
weaknesses
A naive approach in this case would be to 'pad' each message block of length j up to length b and then to encrypt and transmit it. What is important in this case is that blocks with random bit patterns are padded so that the scheme is not vulnerable to cryptographic analyses. The downside of this approach would be an undesirable increase in the data being transmitted because a b length block would have to be transmitted for each j length block (e.g. one character). With character-orientated applications and a cipher with a 64-bit block length, this would amount to an eightfold increase in the transmission bandwidth required.

Ciphertext Feedback
Mode
This kind of increase in required transmission bandwidth can be avoided using a more suitable mode. *Ciphertext Feedback Mode(CFB)* encrypts message blocks of length j bits using a block cipher with a block size of b bits. All message blocks P_i in the following discussion have a length of j bits.

The scheme works with a register. An initialisation vector is written into the register before the first message block is processed. The key is used to encipher the register content before the encryption of each message block. From the value obtained the j higher-valued bits are XORed with the plaintext that is being encrypted. The resulting ciphertext is then written in the j lower-valued bits of the register, which previously was left-shifted by j bits. This feedback of the ciphertext also explains how the name of the scheme originated. Messages are decrypted with the same procedure. The decryption function of the encryption procedure is not required as shown by the following equations in which $S(j, X)$ denotes the higher-valued j bits of X and R_i denotes the content of register R in the encryption or decryption of plaintext block P_i:

$$
\begin{aligned}
R_1 &= IV \\
R_i &= (R_{i-1} \cdot 2^j \bmod 2^b) + C_{i-1} \\
C_i &= S(j, E_K(R_i)) \oplus P_i \\
S(j, E_K(R_i)) \oplus C_i &= S(j, E_K(R_i)) \oplus S(j, E_K(R_i)) \oplus P_i \\
S(j, E_K(R_i)) \oplus C_i &= P_i
\end{aligned}
$$

Figure 3.3 presents an overview of this encryption mode with a second register showing the selection of j higher-valued bits after encryption of the register content. Unlike the one discussed above, this register does not store the status of the procedure and in an actual implementation can just as easily be omitted.

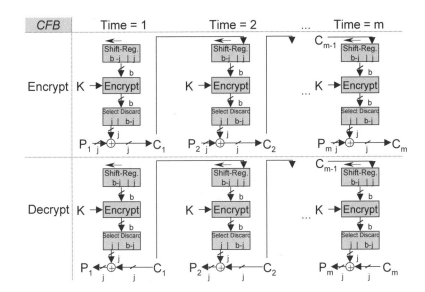

Figure 3.3
Ciphertext Feedback Mode

If a bit error occurs during the transmission of a ciphertext block C_i, a bit error then appears in the same place of corresponding plaintext block P_i'. Furthermore, the error also occurs in the register and produces completely distorted plaintext blocks until it is completely shifted out of the register by subsequent ciphertext blocks. Likewise, if a ciphertext block is lost, the following blocks will be decrypted incorrectly until subsequent ciphertext blocks lead to a resynchronisation of the register content between sender and receiver.

Output Feedback Mode

The concept of encryption using an XOR-operation with a pseudo-random sequence is also realised through the mode called *Output Feedback Mode (OFB)*. In contrast to Ciphertext Feedback Mode, it is not the recovered ciphertext C_i that is fed back into the register but the j higher-valued bits of the enciphered register content. The block cipher in this mode therefore generates a pseudo-random bit sequence, which is completely independent of the plaintext being encrypted. This results in the following relationships (also see Figure 3.4):

$$
\begin{aligned}
R_1 &= IV \\
R_i &= (R_{i-1} \cdot 2^j \bmod 2^b) + S(j, E_K(R_{i-1})) \\
C_i &= S(j, E_K(R_i)) \oplus P_i \\
S(j, E_K(R_i)) \oplus C_i &= S(j, E_K(R_i)) \oplus S(j, E_K(R_i)) \oplus P_i \\
S(j, E_K(R_i)) \oplus C_i &= P_i
\end{aligned}
$$

Figure 3.4
*Output Feedback
Mode*

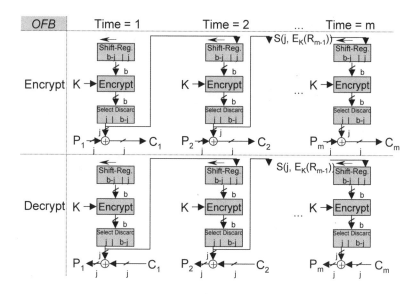

If a bit error occurs between the encryption and decryption of a ciphertext block, then a bit error will also appear in the corresponding place in the plaintext block. Output Feedback Mode, therefore, does not lead to error propagation. However, the loss of one or more message blocks has a disruptive effect on all subsequent blocks and an explicit resynchronisatoin of the procedure is required.

Cryptographically, error propagation is entirely desirable

The characteristic of OFB that it does not propagate errors may be considered an advantage from a data processing view but from a cryptographic standpoint it is usually considered a disadvantage, because it means that attackers can make controllable alterations to texts protected by encryption. This is based on the fact that each modified bit in a ciphertext results in exactly one modified bit in the plaintext after decryption. Even if an attacker cannot know which value the corresponding bit will hold after decryption, he or she can deliberately inverse it. The fact that such an attack can be useful will also be shown in the discussion on the use of security protocols to protect communication in wireless local area networks in Chapter 15.

Multiple use of the same pseudo-random sequence with OFB is not secure!

Another important point to remember with encryption in OFB mode is that the scheme will be completely insecure if the same pseudo-random sequence is used to encrypt two different messages. In a case like this, the XOR-operation of the two ciphertexts produces the XOR-operation of the two corresponding plaintexts, which means that an attacker who knows one of the two plaintexts merely has to compute the pseudo-random sequence and the other plaintext using a simple XOR-operation.

3.2 Data Encryption Standard

The *Data Encryption Standard (DES)* was standardised in the mid-1970s. In 1973 the American National Bureau of Standards, today called the *National Institute of Standards and Technology (NIST)*, had invited proposals for a national encryption standard that would meet the following requirements:

❑ provide a high level of security;

❑ be completely specified and easy to understand;

❑ provide security only on the basis of the secrecy of its keys and not on the basis of the secrecy of the algorithm itself;

❑ potentially be available to all possible users;

❑ be adaptable for use in diverse applications;

❑ be implementable in electronic devices;

❑ be efficient to use;

❑ be capable of being validated;

❑ be exportable.

Requirements of the algorithm

None of the submissions in response to the first call met these requirements, and it was not until the second call that IBM submitted the algorithm 'Lucifer', which turned out to be an encouraging candidate. Lucifer was a symmetric block cipher that worked on 128-bit length blocks and used 128-bit long keys.

IBM's algorithm 'Lucifer' was the only promising submission for the second call

The National Bureau of Standards requested help from the *National Security Agency (NSA)* in assessing the security of the algorithm. The NSA is an authority that also supports the American secret service on cryptographic matters, and it is widely recognised that it has a staff of highly competent cryptographers and mathematicians. The results of the NSA are usually confidential and therefore do not become public knowledge. This makes it difficult to make an exact assessment of the 'cryptographic lead' the NSA enjoys ahead of other cryptographers in the world. That it has had the lead for a long time has been historically proven and the development history of DES also supports this.

Modifications to the algorithm by the NSA

The NSA reduced the block size of the algorithm to 64 bits and its key length to 56 bits. It also made modifications to the internal structure of the substitution tables. These changes were the cause of extensive discussions among cryptographers and it was not until the 1990s that some of these changes could be understood.

Despite all the criticism it attracted, the algorithm was standardised in 1977 [NIS77, NIS88] and released for securing 'non-confidential' government data. Since the algorithm was freely available in the USA, it was also used by private-sector companies and quickly developed into the standard algorithm for encryption. Until the mid-1990s, permission to export DES products from the USA was subject to stringent conditions due to the terms of the US Weapons Act.

Figure 3.5
Overview of the DES algorithm

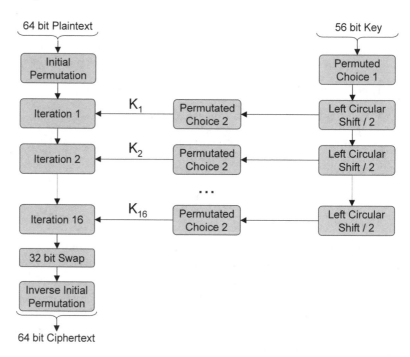

Functioning of DES

Figure 3.5 shows how the DES algorithm is structured. A 64-bit long data block first undergoes an *initial permutation* and is then processed in 16 *iterations*, also called *rounds*. A different *round key* is computed in each round and linked to the data block. After the last iteration, the two 32-bit long halves of the data block are interchanged and the resulting block undergoes a new permutation, which is an inverse operation to the initial permutation. In the illustration, the data block is being worked on in the left half and the key in the right half.

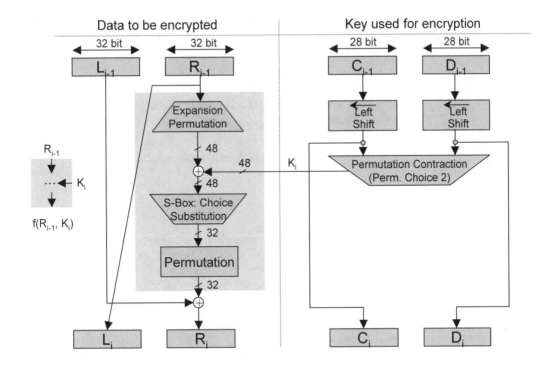

Figure 3.6
One round of the DES algorithm

The internal structure of an iteration step is shown in Figure 3.6, where again one half shows a data block being processed and the other half shows the key. The data block is worked on in two halves with the result of the ith iteration step denoted in the following as L_i and R_i. The left half L_i simply receives the content of the right half R_{i-1} from the previous step. For computing the right half, function $f(R_{i-1}, K_i)$ from the previous step R_{i-1} and the round key K_i are applied to the content of the right half and the resulting value XORed to the content of the left half L_{i-1} of the previous step. The function f first expands and permutes the content of R_{i-1} with the duplication of 16 of the 32 bits, thereby creating a value of the length of 48 bits. This is XORed with the 48-bit long round key and the result of this operation in turn is mapped to a 32-bit long value through eight *substitution boxes (S-boxes)*. During the mapping with the substitution boxes, a 6-bit long value is mapped to a 4-bit long value in each box according to a rule that has been specified for each box. The value obtained according to the substitution boxes is then permuted to obtain the result of function f.

For the computation of round key K_i the round key is also divided into two halves. In each iteration, both halves are shifted

Computation of round key

circularly to the left by one or two bits. The round determines whether the half is shifted one or two positions. After this operation, the 56 bits of the key are mapped to 48 bits to obtain round key K_i.

DES decryption   Decryption with DES uses the same procedure except that the round keys are applied in the reverse. The following computations help to show how the encryption function of DES is reversed. In each step i the two halves of the data block being processed are modified according to the two following equations:

$$
\begin{aligned}
L_i &= R_{i-1} \\
R_i &= L_{i-1} \oplus f(R_{i-1}, K_i)
\end{aligned}
$$

Register content before the first step   For decryption the initial data block $L_0'||R_0'$ before the first iteration step is:

$$
L_0' \,||\, R_0' \;=\; Initial Permutation(ciphertext)
$$

Also:

$$
ciphertext \;=\; Initial Permutation^{-1}(R_{16} \,||\, L_{16})
$$

and therefore overall:

$$
L_0' \,||\, R_0' \;=\; R_{16} \,||\, L_{16}
$$

After one step   After the first iteration step:

$$
\begin{aligned}
L_1' &= R_0' = L_{16} = R_{15} \\
R_1' &= L_0' \oplus f(R_0', K_{16}) = R_{16} \oplus f(R_{15}, K_{16}) \\
&= [L_{15} \oplus f(R_{15}, K_{16})] \oplus f(R_{15}, K_{16}) = L_{15}
\end{aligned}
$$

This relationship continues in all rounds of the decryption because it holds that:

$$
\begin{aligned}
R_{i-1} &= L_i \\
L_{i-1} &= R_i \oplus f(R_{i-1}, K_i) = R_i \oplus f(L_i, K_i)
\end{aligned}
$$

After 16 steps   Therefore after the 16th round it holds that:

$$
L_{16}' \,||\, R_{16}' \;=\; R_0 \,||\, L_0
$$

At the end the two 32-bit long halves of the data block are reversed and the resulting block undergoes inverse initial permutation so that the results of the decryption are:

$InitialPermutation^{-1}(L_0 \parallel R_0) =$

$InitialPermutation^{-1}(InitialPermutation(plaintext)) = plaintext$

In the underlying construction scheme of the DES algorithm, the *Feistel network*
data block being encrypted is divided into two halves and encrypted in several iterations with alternating round keys according to the equations above. This scheme is also called a *Feistel
network* after its creator H. Feistel. Due to its relatively well-
researched cryptographic attributes that have not yet shown any
inherent weaknesses, this structure has been used as the underlying basis for numerous other modern block ciphers in addition to
DES. One of these is the AES candidate *Twofish* [SKW+98].

Independent cryptographers felt that the NSA had weakened
the algorithm by reducing the block size to 64 bits and the key
length to 56 bits . They claimed that the reductions made the al- *Security of DES*
gorithm vulnerable to brute-force attacks, even if at the time these
would have had to be extremely powerful attackers, such as the
secret service of a government. In addition, the NSA modified the
structure of the internal substitution boxes of the algorithm. Because independent cryptographers did not have the opportunity to
evaluate the latter modifications, it was long suspected that these
changes had weakened the algorithm. This meant that the NSA
could have been deciphering messages without ever being in possession of the key.

This suspicion has not ever been confirmed. On the contrary, it *Resistance against*
was obvious in the early 1990s that DES could not be broken by the *differential*
attack technique of differential cryptanalysis [BS90, BS93] pub- *cryptanalysis*
lished in 1990 and the reason was precisely because of the structure of the substitution boxes and the number of iterations carried
out in the algorithm. Sources at the NSA subsequently disclosed
that it had already known about this attack technique in the 1970s
and strengthened DES accordingly.

So it is the short key length of DES that has continued to be its *Key length is main*
main weakness. Even 'average' attackers have been able to carry *weakness*
out brute-force searches through entire key spaces since the mid-
1990s, and this now makes DES an insecure option.

Because it was already possible to predict this potential inade- *Multiple encryption*
quacy of DES in the 1970s through an estimation of technological
development in terms of computing speed, consideration could be
given early enough to the use of multiple encryption with different

keys in order to increase the overall security attainable with the DES algorithm.

Certain algebraic properties can make multiple encryption ineffective

What has to be taken into account, however, is that multiple encryption only results in a longer effective key length if the encryption algorithm does not show certain algebraic properties. If, for example, DES were *closed*, i.e., a third key K_3 exists for all keys K_1, K_2 so that $E(K_2, E(K_1, M)) = E(K_3, M)$ for all messages M, then double encryption using different keys K_1, K_2 would not increase the effective key length. By a complete search of the key space, an attacker in this case would always find a key K_3 that correctly deciphers all messages doubly encrypted with K_1 and K_2.

A similar inadequacy would occur with triple encryption if DES were *pure*, i.e., if a K_4 were to exist for all K_1, K_2, K_3 so that $E(K_3, E(K_2, E(K_1, M))) = E(K_4, M)$ for all messages M. Because DES is neither closed nor pure, multiple encryption for the purpose of extending the effective key length is possible in principle.

Meet-in-the-middle attack with double encryption

Double encryption with two different keys is still insecure because it can be broken with a *meet-in-the-middle* attack. Although the effective key length with DES in this case is 112 bits, *double DES* can be broken with an effort of 2^{56}. All an attacker needs are a few plaintext-ciphertext pairs $C = E(K_2, E(K_1, P))$ that were encrypted with the same key $K = K_1 \| K_2$. What the attacker is 'interested' in is the value X, which always results 'in the middle' between the two encryption steps, so that:

$$X \quad = \quad E(K_1, P) \quad = \quad D(K_2, C)$$

Construction of two tables

An attacker will take advantage of this fact to construct all possible combinations of type $K_1 \| K_2$ keys, with K_1 and K_2 initially unknown. First two tables $T_1[0, 2^{56}]$ and $T_2[0, 2^{56}]$ are set up, with the first one containing the values of $X = E(K_1, P)$ for all possible values of K_1 and the second one containing the values of $X = D(K_2, C)$ for all possible values of K_2. Then both tables are sorted according to the values of X and compared against another value-by-value. A possible candidate for the tuple (K_1, K_2) is found for all matching values of X.

A DES plaintext is always mapped to one of 2^{64} possible ciphertexts due to its block size of 64 bits. Therefore, with 2^{112} possible values for (K_1, K_2), on average $2^{112}/2^{64} = 2^{48}$ 'false alarms' will occur after the first plaintext-ciphertext pair. With each further pair for which both tables have been constructed the probability of such a falsely assumed key value reduces by a factor of 2^{64}. As a result, the probability of a false key after the second pair reduces to 2^{-16}.

Overall this means that the average effort required for the cryptanalysis of a double DES is of the magnitude of 2^{56}, which only represents a minimal increase in security compared to a simple DES, the compromising of which requires an average effort of 2^{55}. This improvement is, therefore, well below the anticipated increase of the average effort to 2^{111} that one would expect of a cryptographic scheme with a key length of 112.

The attack method described can be dealt with through the following triple encryption scheme that was proposed by W. Tuchman in 1979:

Triple DES

$$C = E(K_3, D(K_2, E(K_1, P)))$$

The use of the decryption function D in the middle of the scheme enables devices that implement triple encryption to be used together with devices that only carry out simple encryption. In this case, the three keys are set to the same value (with the resulting security again only being the same as that of simple encryption).

Triple encryption can be used either with two ($K_1 = K_3$) or with three different keys, in which case three different keys are given explicit preference. Until now there has been no publicity about viable attacks on this scheme and so it is considered secure. However, the drawback of multiple encryption compared with simple encryption is that performance is reduced by a third. Consequently, it is more beneficial to use an encryption algorithm that already offers a longer key length for simple encryption to start with. The following section describes the efforts that have gone into standardising such a scheme.

Main drawback of multiple encryption is reduced performance

3.3 Advanced Encryption Standard

In January 1997 the National Institute of Standards and Technology (NIST) in the USA made an official announcement of its plans to develop an *Advanced Encryption Standard (AES)*. As with DES earlier, its principle objective was to develop an encryption standard that could be used to protect sensitive government data. And again, as with DES, it was assumed that as a standard AES would also find wide acceptance in the private sector.

In September 1997 a formal request was published inviting cryptographers from all over the world to submit algorithms for the AES proposal. The conditions stipulated that the algorithms submitted had to be unclassified, published and available without licensing fees worldwide. The algorithms were also required to use

Worldwide call for submission of algorithms

a symmetric block cipher and support a minimum block size of 128 bits as well as key lengths of 128, 192 and 256 bits.

In August 1998 the first AES conference set up to identify possible candidates was held. NIST subsequently nominated a selection of 15 algorithms as candidates and requested comments on them.

Five candidates selected after the second AES conference

A second AES conference took place in March 1999, and the results of various analytic studies carried out by the professional world of cryptography since the first conference were discussed. After this conference, NIST announced in April 1999 that it had narrowed its selection and that five algorithms were now candidates for the final choice of the AES standard. The names of these algorithms were *MARS, RC6, Rijndael, Serpent* and *Twofish*. Another request was made for an intensive analysis of the five algorithms with attention given not only to purely cryptographic aspects but also to implementation considerations, issues relating to possible patent rights in respect to intellectual property and other overall recommendations.

Selection of Rijndael algorithm as future AES

The third and last AES conference was held in May 2000. After an intensive evaluation of its findings, NIST published the Rijndael algorithm as the official proposal for AES in October 2000.

The main justification for selecting the Rijndael algorithm was that it effectively met the central requirements for a secure, efficient, easy to use and flexible algorithm. Even though all five candidates still under consideration during the last round were evenly matched as far as meeting security requirements on the basis of the known level of technology, Rijndael was particularly impressive because of some of its other attributes. It shows good performance in hardware and in software implementations, requires minimal effort in key preparation (thus supporting fast key change), needs minimal memory and uses operations that are relatively simple to protect from timing and power attacks [1].

Publication of AES standard

A first version of the standard was published in February 2001 and allowed a three-month comment period. After a further revision, AES was officially adopted as a standard on November 26, 2001 [NIS01], and came into effect on May 26, 2002.

Parameterisation of Rijndael

The Rijndael algorithm can be used with different block sizes and key lengths. As part of the AES standardisation the block

---

[1]With *timing attacks* an attacker uses the timely behaviour of a cryptographic algorithm to obtain feedback on the current key being used, whereas with *power attacks* similar conclusions are reached on the basis of power demand.

```
// Algorithm: Rijndael_Encrypt
// Input:     in = 16 octets of plaintext (Nb = 4)
//               w = the key prepared for
//                   No. + 1 rounds
// Output:    out = 16 octets of ciphertext

void Rijndael_Encrypt(byte in[4*Nb], byte out[4*Nb],
                      word w[Nb*(Nr+1)])
{ byte state[4,Nb];

    state = in; // Copying of plaintext
                // to state matrix

  AddRoundKey(state, w[0, Nb-1]);

  for (round = 1;  round <= Nr - 1; round++)
  { SubBytes(state);
    ShiftRows(state);
    MixColumns(state);
    AddRoundKey(state, w[round*Nb, (round+1)*Nb-1]);
  }

  SubBytes(state);
  ShiftRows(state);
  AddRoundKey(state, w[Nr*Nb, (Nr+1)*Nb-1]);

  out = state; // Copying of state matrix
               // to output
}
```

Figure 3.7
Overview of Rijndael encryption

size was specified as 128 bits even though the algorithm description in [NIS01] still reflects the flexibility of the algorithm (compare Figure 3.7), giving the block size as parameter Nb in 32-bit words($Nb = 4$). The values 128, 192 and 256 bits are specified for the key length. Depending on the key length, the algorithm iterates internally in 10,12 or 14 rounds (parameter Nr), with a negligible difference between the last round and the other rounds.

The encryption function of Rijndael operates on a state matrix of four-times four octets into which the plaintext being encrypted is initially written and is based on four different operations. In each round, an octet substitution is first carried out on the basis of an S-box *(SubBytes)* and then the columns of the state matrix are interchanged according to line-specific rules *(ShiftRow)*. The

Functioning of Rijndael

subsequent *MixColumns operation* executes a matrix multiplication using a specifically defined matrix with the octets interpreted as elements of the field $GF(2^8)$ and not as numbers. Mathematically, this results in a polynominal multiplication carried out modulo $x^4 + 1$ with the fixed polynominal

$$c(x) = \text{'}03\text{'} \cdot x^3 + \text{'}01\text{'} \cdot x^2 + \text{'}01\text{'} \cdot x + \text{'}02\text{'}$$

The inverse polynominal (modulo $x^4 + 1$) required for the decryption function is:

$$c^{-1}(x) = \text{'}0b\text{'} \cdot x^3 + \text{'}0d\text{'} \cdot x^2 + \text{'}09\text{'} \cdot x + \text{'}0e\text{'}$$

However, a MixColumn operation is not carried out in the last round, which only executes three operations.

The last operation in each round is an XOR-operation with a round-specific key *(AddRoundKey)*. The round key is computed beforehand in an initialising operation before actual encryption (not explained further here).

Decryption function of Rijndael
The decryption function of Rijndael can be realised in two different ways. The variant shown in Figure 3.8 uses the same key schedule for the round key as encryption does but computes the inverse functions for the operations described above in a different sequence.

However, a more efficient variant of the decryption function is one that uses a modified key schedule to realise the sequence of inverse operations analogous to the encryption function.

Proof that the encryption and decryption functions of Rijndael are complementary to one another is not as readily evident as with the DES algorithm. Rijndael clearly uses more complex operations and does not only follow a Feistel structure. This book therefore will not discuss this standard further and interested readers are referred to the AES Standard [NIS01].

3.4 RC4 Algorithm

The RC4 algorithm realises a stream cipher and was created by R. Rivest in 1987. The operations of the algorithm were not published until 1994 when an anonymous announcement appeared on a mailing list.

Encryption in OFB mode
RC4 operates in Output Feedback Mode (OFB), i.e., a pseudo-random bit sequence $RC4(IV, K)$, which is dependent only on the key K being used and an initialisation vector IV, is generated

```
// Algorithm: Rijndael_Decrypt
// Input:      in  = 16 octets ciphertext
//             w   = the key prepared for
//                   No. + 1 rounds
// Output:     out = 16 octets plaintext

void Rijndael_Decrypt(byte in[4*Nb], byte out[4*Nb],
                      word w[Nb*(No.+1)])
{ byte state[4,Nb];

  state = in; // Copying of ciphertext
              // to state matrix

  AddRoundKey(state, w[Nr*Nb, (Nr+1)*Nb-1]);

  for (round = Nr - 1;  round >= 1; round--)
  { InvShiftRows(state);
    InvSubBytes(state);
    AddRoundKey(state, w[round*Nb, (round+1)*Nb-1]);
    InvMixColumns(state);
  }

  InvShiftRows(state);
  InvSubBytes(state);
  AddRoundKey(state, w[0, Nb-1]);

  out = state; // Copy state matrix
               // to the output
}
```

through the algorithm and XORed to the plaintext P being encrypted. Decryption is then realised through an XOR-operation of the ciphertext with the same pseudo-random bit sequence:

$$C = P \oplus RC4(IV, K)$$
$$P = C \oplus RC4(IV, K)$$

The pseudo-random bit sequence is often also referred to as a *keystream*.

For the security of the algorithm it is essential that a keystream used to encrypt a plaintext P_1 is never used to encrypt another plaintext $P_2 \neq P_1$. The initialisation vector IV is therefore also combined with key K, thus ensuring that two different plaintexts are never encrypted with the same initialisation vector $IV_1 = IV_2$. If this condition is violated, an XOR-operation of

*Initialisation vectors
must never be reused!*

the two ciphertexts C_1 and C_2 can compute the XOR-operation of the two plaintexts P_1 and P_2. If $IV_1 = IV_2$, then:

$$C_1 \oplus C_2 = P_1 \oplus RC4(IV_1, K) \oplus P_2 \oplus RC4(IV_2, K)$$
$$= P_1 \oplus RC4(IV_1, K) \oplus RC4(IV_1, K) \oplus P_2$$
$$= P_1 \oplus P_2$$

Known-plaintext attack

If the attacker in this case knows one of the two plaintexts, e.g., P_1, then he or she can use it to compute the other plaintext (known-plaintext attack). This possibility was not taken into account in the standardisation of wireless local networks in the IEEE 802.11 standard. Therefore, the security protocol defined in the standard has serious shortcomings in security (compare Chapter 15).

Figure 3.9
Initialisation of RC4 algorithm

```
// Algorithm: RC4_Init
// Input:     key = the key of the length
//                  KeyAndIvLen already linked to an
//                  initialisation vector that can
//                  only be used once
//
// Ouput:     i, n = 2 state indices modulo 256
//                s = state field with 256 entries

void RC4_Init(byte* key; int& i, n; byte s[256])
{ unsigned int keyIndex, stateIndex;
  byte a;

  // Padding s[] with 0 up to 255
  for (i = 0; i < 256; i++) S[i] = i;

  // Compute initial state field s
  keyIndex = 0; stateIndex = 0;
  for (i=0; i<256; i++)
  { stateIndex = stateIndex + key[keyIndex] + a;
    stateIndex = stateIndex & 0xff; // modulo 256
    a = s[i];
    s[i] = s[stateIndex];
    s[stateIndex] = a;
    if (++keyIndex >= KeyAndIvLen) keyIndex = 0;
  }
  i = 0; n = 0; // Initialise state indices
}
```

The RC4 algorithm can be operated with different length keys and initialisation vectors, with the maximum overall length being 2048 bits. The algorithm operates on a state field $s[\ ]$ with 256 entries and two indices i and n for this field. These variables are initialised in a preliminary step using the initialisation vector and the key (compare Figure 3.9).

```
// Algorithm:  RC4_Process
// Input:         in = the plaintext or ciphertext
//                      being processed
//               len = length of inut
// In/output: i, n = 2 state indices modulo 256
//               s = state field with 256
//                      entries
// Output:       out = the resulting ciphertext
//                      or plaintext

void RC4_Process(byte* in, int len; int& i, n;
                 byte s[256], byte* out)
{ int  j;
  byte a, b;

  for(j = 0; j < len; j++)
  { i = (i+1) & 0xff; // Addition modulo 256
    a = s[i];
    n = (n+a) & 0xff; // Addition modulo 256
    b = s[n];
    s[i] = b;
    s[n] = a;
    out[j] = XOR(in[j], s[((a+b) & 0xff)]);
  }
}
```

Figure 3.10
Encryption or decryption with the RC4 algorithm

Figure 3.10 illustrates how RC4 computes a pseudo-random bit sequence and is used in encryption and decryption.

In terms of the security of RC4, it should be noted that the variable key length of up to 2048 bits means that the algorithm can be operated so that it is secure from brute-force attacks that use resources available in our universe and are based on known technologies (possible future technologies such as quantum informatics explicitly excluded). However, a significant reduction in key length, e.g., to 40 bits, can make operation of the algorithm insecure. For example, the standard specification for keys in the SSL protocol is 40 bits, which *de facto* provides no security (also see Chapter 10). For a long time RC4 with 40 bit long keys had special export status, which is evident by its wide use in diverse software products, in particular WWW browsers. As mentioned earlier, this

Security of RC4

Encryption with 40 bit long keys offers no security!

does not guarantee that information will be secure from spoofing and can just as well be dispensed with. Fortunately, the situation has changed as a result of a relaxing of export rules in the USA and products that use RC4 with 128-bit long keys are now allowed to be sold in most countries.

The company *RSA Data Security, Inc.* claims that RC4 is immune to differential and linear cryptanalysis and that the pseudorandom bit generator has no small cycles. So far no cryptographer has been able to refute this claim.

Improper key scheduling affects security

The key weakness discovered so far in the RC4 algorithm is that, depending on the method used to combine the key and the initialisation vector, the key can be discovered using known plaintext–ciphertext pairs. It is sufficient for the attacker to know the first two octets of the encrypted plaintexts. Depending on the method used to generate the initialisation vectors, the attacker needs around one million or four million such pairs (P_i, C_i), encrypted with the same key K and different initialisation vectors IV_i. If the initialisation vector and the key are merely concatenated before the initialisation routine shown in Figure 3.9 is called up, then there is a certain probability that certain bits of the key can be found out from the first two octets of the plaintext and the ciphertext. This attack technique was presented at a conference in August 2001 [FMS01].

Potential countermeasure

A commentary by R. Rivest describes how to counter this shortcoming through careful preparation of the key with the initialisation vector [Riv01]. Chapter 15 will show that this shortcoming is not just of theoretical interest.

3.5 Summary

At an initial level, symmetric encryption methods can be divided into *block ciphers*, which operate on blocks of a fixed length, and into *stream ciphers*, which in principle encrypt each bit individually.

Encryption modes

A range of different encryption modes exists for block ciphers with the main objective being to ensure that identical plaintext blocks are not mapped continuously to identical ciphertext blocks. *Cipher Block Chaining Mode (CBC)* and *Output Feedback Mode (OFB)* are the two most frequently used modes.

DES

The DES algorithm that was standardised in the 1970s realises a symmetric block cipher with a block length of 64 bits and a key length of 56 bits. Even though no attack method more effective than brute-force has been discovered in the more than 25 years since it was introduced, since the mid-1990s the algorithm has been considered insecure due to its short key length. The now frequently used variant of *Triple-DES (3DES)* encryption with a selectable 112 or 168 bit effective key length displays relatively poor performance behaviour due to the required computation of three iterations with 16 rounds each.

In the mid-1990s development therefore began on a successor to the standard, the *Advanced Encryption Standard (AES)*. Numerous submissions were made from around the world, and after an international public review process the *Rijndael* algorithm was selected. The new AES algorithm implements a symmetric block cipher with a block length of 128 bits and a variable key length of 128, 192 or 256 bits.

AES

The symmetric stream cipher *RC4* is implemented through a pseudo-random bit generator the output of which is XORed (OFB mode) to the plaintext or ciphertext being worked on. RC4 can operate with variable key lengths of up to 2048 bits. It should be noted, however, that before each plaintext is encrypted the key first has to be linked to an initialisation vector in order to ensure that two different plaintexts are not encrypted with the same output stream. Depending on how this key is prepared, the scheme can be insecure. If the key is prepared 'carelessly', it can be computed from around one or four million plaintext-ciphertext pairs.

RC4

3.6 Supplemental Reading

[FMS01] FLUHRER, S.; MANTIN, I.; SHAMIR, A.: Weaknesses in the Key Scheduling Algorithm of RC4. In: *Selected Areas in Cryptography, Lecture Notes in Computer Science* Vol. 2259, Springer, 2001, pp. 1–24

[NIS77] NIST (NATIONAL INSTITUTE OF STANDARDS AND TECHNOLOGY). *FIPS (Federal Information Processing Standard) Publication 46: Data Encryption Standard.* 1977

[NIS01] NIST (NATIONAL INSTITUTE OF STANDARDS AND TECHNOLOGY). *FIPS (Federal Information Processing Standard) Publication 197: Specification for the Advanced Encryption Standard (AES).* 2001

[Riv01]   RIVEST, R.: *RSA Security Response to Weaknesses in Key Scheduling Algorithm of RC4*. 2001. – `http:// www.rsa.com/rsalabs/technotes/wep.html`

3.7   Questions

1.   Is it possible to produce a code dictionary with an encryption algorithm that is operated in ECB mode and has a block length of 128 bits?

2.   Is there a significant cryptographic advantage to keeping an initialisation vector (if it is used) secret?

3.   What is the difference between Ciphertext Feedback Mode and Output Feedback Mode? How does this affect the characteristics of error propagation and synchronisation?

4.   Assume that a block cipher has a 56-bit key length and a 128-bit block length: Can a 'lengthening' of the effective key length using double encryption also be broken in this case through a meet-in-the-middle attack with a still 'justifiable' use of resources? What is the probability in this case of a tuple (K_1, K_2) falsely being assumed to be a key after the nth plaintext-ciphertext pair?

5.   What is the main drawback of the DES algorithm?

6.   Why does double encryption with two different keys not produce an exponential increase of the search space in the cryptanalysis?

7.   Why is the AES algorithm faster than 3DES?

8.   Why can the effective key length with RC4 simply be adjusted?

9.   How can reuse of keystream occur when deploying RC4?

4 Asymmetric Cryptography

This chapter introduces the fundamentals of asymmetric cryptography. It starts with a brief explanation of the basic concept and two main applications of the scheme followed by some preliminary mathematical background needed to understand the algorithms. The chapter continues with an explanation of the three central algorithms by Rivest, Shamir and Adleman (RSA), Diffie and Hellman, and ElGamal. The concluding section gives a short perspective on the use of these algorithms with a specific mathematical field, the elliptic curve, which offers more efficient computation based on shorter keys while maintaining a comparable degree of security.

4.1 Basic Idea of Asymmetric Cryptography

The central idea behind asymmetric cryptography is the use of a different key for decryption from the one used for encryption, or the use of a different key to check a signature than the one used to create this signature. It should also be possible for one of these two keys, indicated as $+K$ below, to be made public without the possibility of it being used to calculate the appropriate private key $-K$ or to decrypt a message encrypted with key $+K$. Expressed in more formal terms, it should not be computationally feasible to use random ciphertext $c = E(+K, m)$ and key $+K$ to compute a plaintext message $m = D(-K, c) = D(-K, E(+K, m))$. In particular, this implies that it should not be feasible to compute key $-K$ from key $+K$.

One of the two keys of a pair can be made public

Key $-K$ is normally referred to as a *(private key)* and $+K$ as a *public key* . In contrast to symmetric cryptgraphy where keys are basically agreed between two (or more) entities A and B and consequently referred to as $K_{A,B}$, asymmetric cryptography allows the allocation of exactly one key pair $(-K_A, +K_A)$ to each entity A.

Asymmetric cryptography favourable for key management

This results in important advantages for key management because now only $2 \cdot n$ keys instead of $n \cdot (n-1)/2$ are needed for secure communication between arbitrary entities from a group of n entities. Furthermore, these keys no longer have to be distributed over confidential channels and instead keys $+K_x$ can be made public.

The two main applications of asymmetric cryptography are:

❏ *Encryption:* When an entity B enciphers a message with public key $+K_A$, it can be assured that only A will be able to decrypt it with its private key $-K_A$.

❏ *Signing:* When an entity A encrypts a message with its private key $-K_A$, then all entities using key $+K_A$ can decipher it. However, as only A is in possession of private key $-K_A$, it is the only entity that can encrypt the message. The ciphertext acts *de facto* as a *signature* of the message.

Authenticity of keys is essential

It is essential that all entities have the assurance that they really know the correct public key of $+K_A$, i.e., the public keys must be authentic. Compared with symmetric cryptography, where keys $K_{A,B}$ must be authentic and distributed over a confidential channel, with asymmetric cryptography the property of authenticity suffices.

Asymmetric methods enable electronic signatures

Another important difference compared with symmetric cryptography is that signatures created using asymmetric methods can also be checked by independent third parties. When signatures are created with a symmetric key $K_{A,B}$, an independent third party is, in principle, not able to distinguish which of the two entities A or B created the signature since both entities know key $K_{A,B}$. In contrast, asymmetric schemes permit the unique allocation of a signature to owner A of key $-K_A$ that was used to produce the signature. Electronic signatures can be realised on this basis.

First publication on asymmetric cryptography

The basic concept of asymmetric schemes described so far along with their possible applications was first published in 1976 by Diffie and Hellman in the article 'New Directions in Cryptography', which was considered a scientific milestone at the time [DH76]. In addition to describing the fundamental idea behind cryptography, this article defined a mathematical algorithm that could be used by two entities to negotiate a shared secret over a non-secured channel. The algorithm was named *Diffie–Hellman key exchange* after its creators. However, this algorithm only implements key exchange and is not an asymmetric cryptographic scheme with any of the attributes listed earlier or deployable for encryption and decryption or signing and signature checking. So although the article made an announcement about asymmetric cryptography and described its properties, it was not able to provide a concrete algorithm to meet its requirements.

The difficulty in developing such a scheme lies in finding an algorithm for encryption and decryption as well as a procedure for the construction of matching key pairs $(+K, -K)$. The scheme needs to ensure that it is impossible to decrypt an encrypted message $E(+K, m)$ if $+K$ is known. The scheme should also meet a number of relevant practical requirements:

Requirements of asymmetric cryptography

- ❏ the key length should be manageable;

- ❏ encrypted messages should not be arbitrarily longer than non-encrypted ones, even if a small constant factor is acceptable;

- ❏ encryption and decryption should not consume too many resources (time, memory).

After the article [DH76] was published, a number of scientists worked on constructing a practical algorithm that would meet the requirements described above. The basic idea was to use a problem familiar from the fields of computer science or mathematics that would be 'difficult' — only with an exponential effort — to solve if only $+K$ were known and 'simple' to solve (with a polynominal effort) if $-K$ were known. The following three problems particularly came to mind:

Construction of concrete algorithms

- ❏ The *knapsack problem*, familiar from computer science, consists of ideally packing a knapsack of a specific capacity, i.e., filled to the maximum, with objects of a specified amount. Merkle, in particular, researched asymmetric schemes based on this problem, but all of them proved to be insecure and could not, therefore, be considered viable from a practical standpoint.

Knapsack problem

- ❏ The *integer-factorisation* problem, which consists of factorising a large integer into its prime factors. This problem forms the basis of the *RSA algorithm*, named after its creators Rivest, Shamir and Adleman. This was the first secure asymmetric algorithm invented and it still represents one of the most important asymmetric algorithms as it is in wide practical use.

Factorisation problem

- ❏ The *problem of the discrete logarithm*, which involves finding the logarithm of a number for a given basis in finite fields,

Problem of the discrete logarithm

is the cornerstone of the key exchange algorithm by Diffie–Hellman as well as for the asymmetric cryptographic scheme according to ElGamal, both of which still have great practical significance today.

How these schemes function will be explained in the following sections. First an understanding of some mathematical principles is required.

4.2 Mathematical Principles

This section introduces the mathematical principles of asymmetric cryptography. Most of the relationships presented build on one other so that no previous mathematical knowledge is required other than what is normally taught in school. Readers who have a sound mathematical background may find the explanations somewhat detailed and are requested to have some patience.

Divisibility   In the following, let $\mathbb{Z}$ be the number of positive and negative integers and $a, b, k, n \in \mathbb{Z}$. We say a *divides* b (Notation: '$a|b$') when an integer $k \in \mathbb{Z}$ exists so that $a \cdot k = b$. A positive integer a is called *prime* when its only divisors are 1 and a.

Remainder with   Moreover, r *is the remainder of* a *divided by* n when $r = a -$
division $\lfloor a/n \rfloor \cdot n$, with $\lfloor x \rfloor$ being the largest integer that is smaller than or equal to x. For example, 4 is the remainder of 11 divided by 7, since $4 = 11 - \lfloor 11/7 \rfloor \cdot 7$. This relationship can also be written in another way using $a = q \cdot n + r$, with $q = \lfloor a/n \rfloor$. For the remainder r of the division of a by n we also write $a \text{ MOD } n$.

Congruence modulo   We say b is *congruent* a *modulo* n when b has the same re-
of a number   mainder as a when divided by n. n therefore divides the difference $(a - b)$ and we write $b \equiv a \mod n$. Particular attention should be paid here to the different way in which MOD and $\mod$ are written. Whereas $a \text{ MOD } b$ denotes a concrete number, namely the remainder that results when a is divided by b, the style $\mod n$ is a more detailed specification of a congruence and to an extent represents a rounding off of character $\equiv$. Without this detail there would be no clear reference to the number to which the two other numbers are actually congruent.

Examples:

$$4 \equiv 11 \mod 7, \qquad 25 \equiv 11 \mod 7, \qquad 11 \equiv 25 \mod 7,$$
$$11 \equiv 4 \mod 7, \qquad -10 \equiv 4 \mod 7$$

Residue class   Because the remainder r of division by n is always smaller than n, we often represent the set $\{x \text{ MOD } n \mid\mid x \in \mathbb{Z}\}$ through its elements $\{0, 1, \ldots, n-1\}$. Each element a of this set always stands for an entire set of numbers that all produce the same remainder a when divided by n. These sets are also called *residue classes*. Normally they are noted as $[a]_n := \{x \mid\mid x \in \mathbb{Z} : x \equiv a \mod n\}$ and their combined set as $\mathbb{Z}_n := \{[a]_n \mid\mid a \in \mathbb{Z}\}$.

The study of remainders with division by a number leads to *modular arithmetic*; some of its properties and computing rules are summarised in Table 4.1.

Definition 4.1 *The number c is called the **greatest common divisor (GCD)** of two numbers a and b when c divides both numbers a and b and c also is the greatest integer with this property. Therefore, each other common divisor d of a and b is also a divisor of c. The greatest common divisor of a and 0 is defined for all $a \in \mathbb{Z}$ as $|a|$, the absolute value of a.*

Characteristic	Expression
Commutative laws	$(a + b) \text{ MOD } n = (b + a) \text{ MOD } n$
	$(a \cdot b) \text{ MOD } n = (b \cdot a) \text{ MOD } n$
Associative laws	$[(a + b) + c] \text{ MOD } n = [a + (b + c)] \text{ MOD } n$
	$[(a \cdot b) \cdot c] \text{ MOD } n = [a \cdot (b \cdot c)] \text{ MOD } n$
Distributive law	$[a \cdot (b + c)] \text{ MOD } n = [(a \cdot b) + (a \cdot c)] \text{ MOD } n$
Identities	$(0 + a) \text{ MOD } n = a \text{ MOD } n$
	$(1 \cdot a) \text{ MOD } n = a \text{ MOD } n$
Inverses	$\forall a \in \mathbb{Z}_n : \exists (-a) \in \mathbb{Z}_n : a + (-a) \equiv 0 \mod n$
	$p \text{ is prime} \Rightarrow \forall a \in \mathbb{Z}_p : \exists (a^{-1}) \in \mathbb{Z}_p : a \cdot (a^{-1}) \equiv 1 \mod p$

Table 4.1
Properties of modular arithmetic

The greatest common divisor of a and b is usually noted as $gcd(a, b)$. The above definition can also be formulated in an abbreviated style:

$$c = gcd(a, b) :\Leftrightarrow (c \mid a) \wedge (c \mid b) \wedge [\forall d : (d \mid a) \wedge (d \mid b) \Rightarrow (d \mid c)]$$

with $gcd(a, 0) := |a|$

The following theorem can be used as the basis for efficiently computing the greatest common divisor of two numbers:

Theorem 4.1 *For all positive integer numbers a and b the greatest common divisor of a and b is equal to the greatest common divisor of b and a MOD b:*

$$\forall a, b \in \mathbb{Z}^+ : \gcd(a, b) = \gcd(b, a \text{ MOD } b)$$

Proof of this theorem is presented in two arguments:

❏ As the GCD of a and b divides a as well as b, it also divides each linear combination of them, in particular the linear combination $(a - \lfloor a/b \rfloor \cdot b) = a \text{ MOD } b$. Therefore $\gcd(a, b) \mid \gcd(b, a \text{ MOD } b)$.

❏ As the GCD of b and $a \text{ MOD } b$ divides b as well as $a \text{ MOD } b$, it also divides each linear combination of them, in particular also the linear combination $\lfloor a/b \rfloor \cdot b + (a \text{ MOD } b) = a$. Therefore $\gcd(b, a \text{ MOD } b) \mid gcd(a, b)$. □

Recursive computation of GCD

Theorem 4.1 directly leads to the correctness of using the *Euclidean algorithm* to compute the GCD of two numbers, formulated in a notation based on programming languages *C* or *C++* in Figure 4.1.

Figure 4.1
Euclidean algorithm

```
// Algorithm: Euclidean
// Input:     two positive integers a, b
// Ouput:     the greatest common divisor of a, b

int Euclid(int a, b)
{
  if (b = 0) {return(a);}
             {return(Euclid(b, a MOD b));}
}
```

It is often helpful to use an extended form of the Euclidean algorithm along with the greatest common divisor c to compute another two coefficients m and n with c representing the linear combination of a and b so that $c = \gcd(a, b) = m \cdot a + n \cdot b$. This algorithm is shown in Figure 4.2.

Proof by full induction

The correctness proof for the extended Euclidean algorithm is carried out by full induction::

❏ Base case $(a, 0)$: $\gcd(a, 0) = a = 1 \cdot a + 0 \cdot 0$

❏ Induction hypothesis: the extended Euclidean algorithm computes c', m', n' correctly.

❏ Induction of $(b, a \text{ MOD } b)$ to (a, b):

$$c = c' = m' \cdot b + n' \cdot (a \text{ MOD } b)$$
$$= m' \cdot b + n' \cdot (a - \lfloor a/b \rfloor \cdot b)$$
$$= n' \cdot a + (m' - \lfloor a/b \rfloor \cdot n') \cdot b \quad □$$

Figure 4.2
*Extended Euclidean
algorithm*

```
// Algorithm: Extended Euclidean
// Input:     two positive integers a, b
// Output     numbers c, m, n, so that:
//               c = gcd(a, b) = m * a + n * b
// Note:      Floor(x) computes the greatest
//               integer number <= x

struct{int c, m, n}  ExtendedEuclidean(int a, b)
{
  int c', m', n';

  if (b = 0) {return(a, 1, 0); }
  (c', m', n') = ExtendedEuclidean(b, a MOD b);
  (c, m, n)    = (c', n', m' - Floor(a / b) * n');
  return(c, m, n);
}
```

It can also be shown that the runtime of both Euclidean algorithms is of the magnitude of $O(b)$. The proof of this assessment will not be dealt with here and the reader is referred to [CLR90, Section 33.2] for additional information. The discussion of the two Euclidean algorithms can be summarised with the following lemma:

*Runtime of Euclidean
algorithms*

Lemma 4.1 *Let $a, b \in \mathbb{Z}$ and $c = \gcd(a, b)$. We then have numbers $m, n \in \mathbb{Z}$ so that: $c = m \cdot a + n \cdot b$.*

We can use this lemma to prove the following theorem:

Theorem 4.2 *(Euclidean) When a prime number p divides the product of two integers a, b, then it is dividing at least one of the two numbers:*

$$p \mid (a \cdot b) \Rightarrow (p \mid a) \vee (p \mid b)$$

Proof of this theorem by provided by two arguments:

❏ If $p \mid a$, then proof would exist.

❏ If not, then $\gcd(p, a) = 1$ holds.

$$\Rightarrow \exists m, n \in \mathbb{Z} : 1 = m \cdot p + n \cdot a$$
$$\Leftrightarrow b = m \cdot p \cdot b + n \cdot a \cdot b$$

However, as p divides the product $(a \cdot b)$, then p divides both summands of the equation and therefore also divides the sum, which is b. □

Theorem 4.2, proven above, is useful in turn for proof of the *fundamental theorem of arithmetic*:

Fundamental theorem of arithmetic

Theorem 4.3 *The factorisation of an integer is unique up to the sequence of its prime factors.*

For proof of this theorem we will show that each integer with non-unique factorisation has a real divisor with non-unique factorisation. A contradiction occurs if the number is reduced to a prime number due to a repeated application of this argument. As the correct inference is that a false statement cannot be derived from a true assumption, the assumption therefore must be wrong (basic principle of proof through contradiction). Let us assume therefore that n is a number with non-unique factorisation, i.e.:

Proof through contradiction

$$n = p_1 \cdot p_2 \cdot \cdots \cdot p_r$$
$$= q_1 \cdot q_2 \cdot \cdots \cdot q_s$$

The prime factors are not necessarily all different but the second factorisation is not merely a reordering of the first.

Because p_1 divides the number n, it also divides the product $q_1 \cdot q_2 \cdot \ldots \cdot q_s$. A repeated application of Theorem 4.2 results in at least one q_i that is divisible by p_1. However, as q_i is a prime number, p_1 and q_i have to be equal so that we can divide by p_1 and thus arrive at n/p_1, which must be a number with non-unique factorisation. $\square$

Division by factors with congruences

We can use Theorem 4.3 to prove the following corollary that gives the conditions for dividing a factor on both sides of a congruence:

Corollary 4.1 *Let $a, b, c, m \in \mathbb{Z}\backslash\{0\}$, $gcd(c, m) = 1$ and furthermore $(a \cdot c) \equiv (b \cdot c) \mod m$. This implies that $a \equiv b \mod m$.*

For proof: As $(a \cdot c) \equiv (b \cdot c) \mod m$, it follows that the difference between $(a \cdot c)$ and $(b \cdot c)$ has to be divisible by m without remainder, i.e., $\exists n \in \mathbb{Z} : (a \cdot c) - (b \cdot c) = n \cdot m$

$$\Leftrightarrow (a - b) \quad \cdot \quad c \quad = \quad n \quad \cdot \quad m$$
$$\Leftrightarrow \overbrace{p_1 \cdot \ldots \cdot p_i} \cdot \overbrace{q_1 \cdot \ldots \cdot q_j} = \overbrace{r_1 \cdot \ldots \cdot r_k} \cdot \overbrace{s_1 \cdot \ldots \cdot s_l}$$

The second line of this equivalence shows the separate components of the line above factorised into their unique prime factors according to Theorem 4.3. However, the same prime factors have to appear on both sides of the equation (again because of Theorem 4.3). Furthermore, the prime factors of c all differ from the prime factors of m since c and m are assumed to be relatively prime ($gcd(c, m) = 1$). Consequently, $q_1, ..., q_j$ can be successively divided by all of them without one of the prime factors $s_1, ..., s_l$ being eliminated. An equation of the form $a - b = n' \cdot m$ is all that still remains. This in turn means that $(a - b)$ can be divided by m without remainder, which is equivalent to $a \equiv b \mod m$. $\square$

Definition 4.2 *(Euclidean Φ function) Let $\Phi(n)$ denote the number of positive integers $a_1, ..., a_t$ that are smaller than n and relatively prime to n, i.e., $\forall i \in \{1, ..., t\}: (a_i < n) \land \gcd(a_i, n) = 1$.*   *Euclidean Φ function*

The best way to understand this function is to look at a few examples: $\Phi(4) = 2$ (as $\{1, 3\}$ are relatively prime to 4), $\Phi(6) = 2$ (as $\{1, 5\}$ are relatively prime to 6), $\Phi(7) = 6$ (as $\{1, 2, 3, 4, 5, 6\}$ are relatively prime to 7), $\Phi(15) = 8$ (as $\{1, 2, 4, 7, 8, 11, 13, 14\}$ are relatively prime to 15). Basically it holds for all prime numbers p that $\Phi(p) = p - 1$ since prime numbers are only divisible by 1 and themselves.

Even though the review of a set of numbers that are smaller than a number and are relatively prime to this number would not initially appear to be particularly beneficial, it does provide 'raw material' for one of the most important asymmetrical cryptographic schemes, the RSA algorithm. The following central theorem serves as its 'foundation':   *Mathematical principle of RSA algorithm*

Theorem 4.4 *(Euclidean) Let n and b be positive integers and relatively prime to one another, i.e., $\gcd(n, b) = 1$. Then it holds that $b^{\Phi(n)} \equiv 1 \mod n$.*   *Euclidean theorem*

To prove this theorem, we consider $t = \Phi(n)$ and $\{a_1, ..., a_t\}$, the set of numbers that are prime to n. We also define the set $\{r_1, ..., r_t\}$ as the remainders $b \cdot a_i$ MOD n so that it always holds that $b \cdot a_i \equiv r_i \mod n$:

❏ Note that $i \neq j \Rightarrow r_i \neq r_j$. The reason is that a pair (i, j) of indexes could otherwise be found so that $b \cdot a_i \equiv b \cdot a_j \mod n$. However, as $\gcd(n, b) = 1$, with Corollary 4.1 this would imply that $a_i \equiv a_j \mod n$, which is not possible because all a_i are per definition different integers between 0 and n.

❏ We could also deduce that all r_i are relatively prime to n since each common divisor k of r_i and n, i.e., with $n = k \cdot m$ and

$r_i = p_i \cdot k$ for suitable m and p_i, would also be a divisor of a_i:

$$b \cdot a_i \equiv (p_i \cdot k) \mod (k \cdot m)$$
$$\Rightarrow \exists s \in \mathbb{Z}: (b \cdot a_i) - (p_i \cdot k) = s \cdot k \cdot m$$
$$\Leftrightarrow (b \cdot a_i) = s \cdot k \cdot m + (p_i \cdot k)$$

Since k divides each of the two summands on the right side, it also has to divide the left side. As b is assumed to be relatively prime to n, it would consequently have to divide a_i, which would however be a contradiction to the definition of a_i, since a_i per definition is relatively prime to n.

❑ $\{r_1, ..., r_t\}$ is consequently a set of $\Phi(n)$ different integers that are all relatively prime to n. This means that the numbers are exactly the same as those in set $\{a_1, ..., a_t\}$, but normally in a different sequence. This therefore brings us to the conclusion that $r_1 \cdot ... \cdot r_t = a_1 \cdot ... \cdot a_t$.

❑ Let us now study the congruence:

$$r_1 \cdot ... \cdot r_t \equiv b \cdot a_1 \cdot ... \cdot b \cdot a_t \qquad \mod n$$
$$\Leftrightarrow r_1 \cdot ... \cdot r_t \equiv b^t \cdot a_1 \cdot ... \cdot a_t \qquad \mod n$$
$$\Leftrightarrow r_1 \cdot ... \cdot r_t \equiv b^t \cdot r_1 \cdot ... \cdot r_t \qquad \mod n$$

Since all r_i are relatively prime to n, we can apply Corollary 4.1 and divide by the product $r_1 \cdot ... \cdot r_t$ to arrive at $1 \equiv b^t \mod n \Leftrightarrow 1 \equiv b^{\Phi(n)} \mod n$. ☐

Even though we will use Theorem 4.4 to prove the basis of the RSA algorithm, we still need an additional result that can help us to compute the Euclidean Φ function for the product of two numbers. To provide proof, we first derive a much more powerful result, the *Chinese remainder theorem*:

Chinese remainder theorem

Theorem 4.5 *(Chinese remainder theorem) Let $m_1, ..., m_r$ be positive integers that pair-wise are relatively prime to one another, i.e., $\forall\, i \neq j: \gcd(m_i, m_j) = 1$. Let furthermore $a_1, ..., a_r$ be arbitrary integers. Then there exists an integer a so that a fulfils all the following congruences:*

$$a \equiv a_1 \mod m_1$$
$$a \equiv a_2 \mod m_2$$
$$...$$
$$a \equiv a_r \mod m_r$$

Furthermore, a is unique modulo $M := m_1 \cdot ... \cdot m_r$.

To prove this theorem, we first define the number $M_i :=$ $(M/m_i)^{\Phi(m_i)}$ for all $i \in \{1, ..., r\}$.

❏ As M_i is by definition relatively prime to m_i, we can apply Theorem 4.4 and know that $M_i \equiv 1 \mod m_i$.

❏ As M_i is divisible by m_j for all $j \neq i$, we also know that $\forall \, j \neq i : M_i \equiv 0 \mod m_j$.

❏ We can therefore define the solution of simultaneous congruences as:

Constructing the solution of simultaneous congruences

$$a := a_1 \cdot M_1 + a_2 \cdot M_2 + ... + a_r \cdot M_r$$

Based on the two arguments on congruences satisfied by M_i given above, it holds that a actually meets all required congruences of the form $a \equiv a_i \mod m_i$.

❏ To show the uniqueness of this solution modulo M, we will look at an arbitrarily different integer b, which fulfils the r required congruences. As $a \equiv c \mod n$ and $b \equiv c \mod n$ imply that also $a \equiv b \mod n$, it holds that:

Uniqueness of the solution modulo M

$$\forall \, i \in \{1, .., r\} : a \equiv b \mod m_i$$
$$\Rightarrow \forall \, i \in \{1, .., r\} : m_i | (a - b)$$
$$\Rightarrow M | (a - b) \text{ since all } m_i \text{ pair-wise are relatively prime}$$
$$\Leftrightarrow a \equiv b \mod M \quad \square$$

We will now use Theorem 4.5 to prove the following lemma that helps us to compute the Euclidean Φ function for the product of two relatively prime numbers:

Computation of Φ function for the product of relatively prime numbers

Lemma 4.2 *Let $m, n \in \mathbb{Z}^+$ and $\gcd(m, n) = 1$. It then holds that:*

$$\Phi(m \cdot n) = \Phi(m) \cdot \Phi(n)$$

For the proof of the lemma let us consider an arbitrary positive integer a, which is smaller than $(m \cdot n)$ and relatively prime to $(m \cdot n)$. In other words, a is one of the numbers that is counted by $\Phi(m \cdot n)$:

❏ Let us consider the tuple $a \rightarrow (a \text{ MOD } m, a \text{ MOD } n)$. The number a is relatively prime to m and also to n as it would otherwise divide the product $(m \cdot n)$.

This makes $(a \text{ MOD } m)$ relatively prime to m and $(a \text{ MOD } n)$ relatively prime to n, as $a = \lfloor a/m \rfloor \cdot m + (a \text{ MOD } m)$, which means that if a common divisor of m and $(a \bmod m)$ exists, it too would have to divide a.

A tuple $(a \bmod m, a \bmod n)$ therefore corresponds to each number a that is counted by $\Phi(m \cdot n)$, with $(a \bmod m)$ counted by $\Phi(m)$ and $(a \bmod n)$ counted by $\Phi(n)$.

❏ On the basis of the second result of Theorem 4.5, the uniqueness of a solution a modulo $(m \cdot n)$ for the simultaneous congruences:

$$a \equiv (a \bmod m) \quad \bmod m$$
$$a \equiv (a \bmod n) \quad \bmod n$$

we can deduce that distinct integers counted by $\Phi(m \cdot n)$ correspond to different tuples $(a \bmod m, a \bmod n)$.

Another way to understand this is through the assumption that two numbers $a \neq b$ counted by $\Phi(m \cdot n)$ correspond to the same pair $(a \bmod m, a \bmod n)$. However, this assumption creates a contradiction since b would also have to fulfil the two congruences:

$$b \equiv (a \bmod m) \quad \bmod m$$
$$b \equiv (a \bmod n) \quad \bmod n$$

At most $\Phi(m \cdot n)$
counts $\Phi(m) \cdot \Phi(n)$
numbers

According to the second part of Theorem 4.5, the solution of these congruences is uniquely modulo $(m \cdot n)$. Consequently, all corresponding pairs must be different. The number of integers that can be counted by $\Phi(m \cdot n)$ is therefore limited by the number of distinct pairs of numbers that can be counted by $\Phi(m)$ and $\Phi(n)$.

$$\Phi(m \cdot n) \leq \Phi(m) \cdot \Phi(n)$$

❏ Let us now consider a number pair (b, c) in which b is counted by $\Phi(m)$ and c by $\Phi(n)$. Using the first part of Theorem 4.5, we can construct a unique positive integer that is smaller than and relatively prime to $(m \cdot n)$ and fulfils the two following congruences:

$$a \equiv b \quad \bmod m$$
$$a \equiv c \quad \bmod n$$

At most $\Phi(m) \cdot \Phi(n)$
can be $\Phi(m \cdot n)$

However, as per definition of the Euclidean Φ function, a maximum of $\Phi(m \cdot n)$ different numbers can exist that are smaller than and relatively prime to $(m \cdot n)$, the number of pairs of the type $(a \text{ MOD } m, a \text{ MOD } n)$ can at most amount to $\Phi(m \cdot n)$:

$$\Phi(m \cdot n) \geq \Phi(m) \cdot \Phi(n) \quad \square$$

We have used Lemma 4.2 to prove all mathematical principles needed to understand how the RSA algorithm functions. The following section uses this basis to provide a clear overview of the actual algorithm.

4.3 The RSA Algorithm

The RSA algorithm was invented in 1977 by Rivest, Shamir and Adleman and is based on Theorem 4.4. The way it functions is relatively easy to explain.

Let p and q be large prime numbers that are distinct from one another and $n = p \cdot q$. We also assume to know two integers $e, d \in \mathbb{Z}$ so that $d \cdot e \equiv 1 \mod \Phi(n)$ (how such numbers are calculated will be explained later).

Prerequisites of RSA algorithm

Let M be a positive integer that represents an encrypted message and is smaller than and relatively prime to n. If, for example, M should encode a text message, one can specify that letters {A, B, ..., Z} are encoded by numbers $\{10, 11, ..., 35\}$ and the blank by the number 99. The character string 'HELLO' would then be represented by the number 1714212124. If necessary, the number can also be broken up into blocks of smaller numbers: 17142 12124.

Encoding messages as numbers

To encrypt M, one computes:

Encryption

$$E = M^e \text{ MOD } n$$

This computation can be executed efficiently using the *'Square and Multiply'* algorithm, sometimes also called *'Repeated Squaring'* algorithm [CLR90, Section 33.6].

To decrypt encrypted message E again, one computes:

Decryption

$$M' = E^d \text{ MOD } n$$

The correctness of the RSA algorithm can be explained on the basis of the following two observations:

Correctness

❏ As $d \cdot e \equiv 1 \mod \Phi(n) \Rightarrow$

$$\exists k \in \mathbb{Z}: (d \cdot e) - 1 = k \cdot \Phi(n)$$
$$\Leftrightarrow (d \cdot e) = k \cdot \Phi(n) + 1$$

❑ Therefore, it holds that:

$$M' \equiv E^d \equiv M^{(e \cdot d)} \equiv M^{(k \cdot \Phi(n)+1)} \equiv 1^k \cdot M \equiv M \mod n \quad \square$$

With RSA keys can be used in both directions

As $(d \cdot e) = (e \cdot d)$, this operation also works in the opposite direction, i.e., the number d can be used for encryption and the associated e for decryption. This characteristic of the RSA algorithm enables the same key pair to be used for the two following operations:

❑ Receiving messages that have been encrypted with one's own public key

❑ Sending messages that through encryption were signed with one's own private key

Creating RSA keys

The following steps are required to create a key pair that can be used with RSA:

1. Randomly choose two large prime numbers p and q (each with about 100 to 200 decimal points).

2. Compute $n = p \cdot q$ and $\Phi(n) = (p-1) \cdot (q-1)$ (Lemma 4.2).

3. Randomly choose an e and compute using the extended Euclidean algorithm c, d as well as $\gcd(e, \Phi(n))$ so that $e \cdot d + \Phi(n) \cdot c = \gcd(e, \Phi(n))$.

   If $\gcd(e, \Phi(n)) \neq 1$, then choose a new e so long until you find an e that is relatively prime to $\Phi(n)$. Note that this construction ensures that $e \cdot d \equiv 1 \mod \Phi(n)$.

4. The public key is the pair (e, n).

5. The private key is the pair (d, n).

Security of RSA

The security of this scheme is based on the difficulty of factorising the large integer n into its prime factors $p \cdot q$, since it is simple to compute $\Phi(n)$ and thus d if p and q are known.

We will not delve into why it is 'difficult,' i.e., computationally too intensive, to factorise large integers using schemes already known to us, because this would require an extensive discussion on the mathematical relationships relevant in this context. At this juncture we will content ourselves with the somewhat vague statement that when p and q fulfill certain properties, the runtime of the best known algorithms is exponential to the number of binary digits n.

However, we should briefly point out some pitfalls in conjunction with the implementation of RSA so the reader has a certain basic awareness of the dangers that exist for those who are 'inexperienced' at working with the relevant mathematical relationships:

Potential pitfalls when implementing RSA

❏ For example, if p and q are chosen in an 'unfortunate' way, algorithms might exist that can factorise n more efficiently than in exponential runtime. RSA encryption that uses keys constructed this way will therefore not be secure.

❏ Therefore, p and q should be about the same bit length and be sufficiently large. Likewise, the difference $(p - q)$ should not be too small.

❏ Should a small encryption exponent be selected for the public key (which enables operations with a public key to be executed noticeably more quickly), then there are other constraints to be considered, for example:

$$\gcd(p - 1, 3) = 1 \text{ and } \gcd(q - 1, 3) = 1$$

❏ The security of RSA strongly depends on whether the prime numbers p and q are truly random because otherwise it is easy for attackers to imitate this process and guess the factorisation of n relatively quickly. It goes without saying that all cryptographic schemes should have a 'sufficiently random' key creation process.

We have one concluding comment to make in regard to the security of RSA implementations. It is our recommendation that users either work with an existing and widely used implementation or, if it is their own implementation, make sure to have it checked by an appropriately experienced mathematician or, better yet, a cryptographer.

RSA implementations should be checked by cryptographers

4.4 The Problem of the Discrete Logarithm

Section 4.1 mentioned that asymmetric cryptography can be implemented on the basis of the factorisation of large integers as well as the *problem of the discrete logarithm*. This section deals with the mathematical principles relevant to the problem of the discrete logarithm.

Finite group

Definition 4.3 *(Finite group) A* **group** $(S, \oplus)$ *is a set S together with a binary operation $\oplus$ for which the following properties hold:*

❑ **Closure:** $\forall\, a, b \in S\colon\ a \oplus b \in S$

❑ **Identity:** $\exists e \in S\colon\ \forall\, a \in S\colon\ e \oplus a = a \oplus e = a$

❑ **Associativity:** $\forall\, a, b, c, \in S\colon\ (a \oplus b) \oplus c = a \oplus (b \oplus c)$

❑ **Inverse element:** $\forall\, a \in S\colon \exists b \in S\colon\ a \oplus b = b \oplus a = e$

If a group $(S, \oplus)$ also satisfies the commutative law $\forall\, a, b \in S\colon$ $a \oplus b = b \oplus a$, then it is called an **Abelian group.** *If a group $(S, \oplus)$ only has a finite set of elements, i.e., $|S| < \infty$, then it is called a* **finite group.**

Examples

The following examples clarify this further:

$(\mathbb{Z}_n, +_n)$

❑ Group $(\mathbb{Z}_n, +_n)$ of the residue classes modulo n with the operation Addition modulo n, which is defined as follows:

    ❑ $\mathbb{Z}_n := \{[0]_n, [1]_n, ..., [n-1]_n\}$

    ❑ $[a]_n := \{b \in \mathbb{Z} \mid b \equiv a \mod n\}$

    ❑ $+_n$ is defined so that $[a]_n +_n [b]_n = [a+b]_n$

is a finite Abelian group. Direct proof can be calculated using the computing rules of modular arithmetic (also see Table 4.1).

$(\mathbb{Z}_n^*, \cdot_n)$

❑ Group $(\mathbb{Z}_n^*, \cdot_n)$ of the residue classes modulo n with the operation Multiplication modulo n, which is defined as follows:

    ❑ $\mathbb{Z}_n^* := \{[a]_n \in \mathbb{Z}_n \mid \gcd(a, n) = 1\}$

    ❑ $\cdot_n$ is defined so that $[a]_n \cdot_n [b]_n = [a \cdot b]_n$

is a finite Abelian group. Direct proof can be calculated here too using the computing rules of modular arithmetic.

Note that $\mathbb{Z}_n^*$ only contains those elements from $\mathbb{Z}_n$ for which inverse elements modulo n exist. Z_{15}^* is presented here explicitly as an example:

$$\mathbb{Z}_{15}^* = \{[1]_{15}, [2]_{15}, [4]_{15}, [7]_{15}, [8]_{15}, [11]_{15}, [13]_{15}, [14]_{15}\}, \text{ as}$$

$$1 \cdot 1 \equiv 1 \mod 15, \qquad 2 \cdot 8 \equiv 1 \mod 15,$$
$$4 \cdot 4 \equiv 1 \mod 15, \qquad 7 \cdot 13 \equiv 1 \mod 15,$$
$$11 \cdot 11 \equiv 1 \mod 15, \qquad 14 \cdot 14 \equiv 1 \mod 15$$

When it is clear from the context that group $(\mathbb{Z}_n, +_n)$ or $(\mathbb{Z}_n^*, \cdot_n)$ is meant, then the equivalence classes $[a]_n$ are often represented by their elements a and the operations '$+_n$' and '$\cdot_n$' are noted with '$+$' or '$\cdot$'.

Equivalence classes represented by elements

Definition 4.4 *(Finite fields) A* **field** *$(S, \oplus, \odot)$ is a set S together with two binary operations $\oplus$ and $\odot$ so that the following holds:*

Finite fields

❏ *$(S, \oplus)$ and $(S \backslash \{e_\oplus\}, \odot)$ are Abelian groups, i.e., in regard to the addition it is only the neutral element that does not need to have an inverse element in regard to $\odot$ in S, and*

❏ *$\forall\, a, b, c \in S$: $a \odot (b \oplus c) = (a \odot b) \oplus (a \odot c)$, i.e., both operations must satisfy a* **distributive law.**

If a field $(S, \oplus, \odot)$ only has a finite set of elements, i.e., $|S| < \infty$, it is also called a **finite field.**

Thus $(\mathbb{Z}_p, +_p, \cdot_p)$ is a finite field for all prime numbers p, which is easily proven by the computing rules of modular arithmetic.

Definition 4.5 *(Primitive root, generator) Let $(S, \circ)$ be a group $g \in S$ and $g^a := g \circ g \circ \ldots \circ g$, i.e., g is linked a-times with itself, $a \in \mathbb{Z}^+$. Then g is called a* **primitive root** *or a* **generator** *of $(S, \circ)$ precisely when:*

Primitive root

$$\{g^a \mid 1 \le a \le |S|\} = S$$

1 is therefore a primitive root of $(\mathbb{Z}_n, +_n)$ and 3 is a primitive root of $(\mathbb{Z}_7^*, \cdot_7)$. Not all groups have primitive roots, and the groups for which a primitive root exists are also called *cyclic groups.*

For the additive groups of the residue classes modulo n the neutral element in respect to the multiplication 1 can be specified as the primitive root. A generator does not always exist for multiplicative groups $(\mathbb{Z}_n^*, \cdot_n)$. This raises the question of which circumstances make a multiplicative group $(\mathbb{Z}_n^*, \cdot_n)$ cyclic.

When is a multiplicative group $(\mathbb{Z}_n^*, \cdot_n)$ considered cyclic?

The following theorem gives the answer to this question. However, in contrast to the results discussed in the preceding sections for which it was relatively easy to provide proof, this theorem as well as others in this section do not lend themselves to brief coverage. Complete explanations will therefore no longer be provided and the reader is instead referred to the appropriate literature.

Theorem 4.6 *Group $(\mathbb{Z}_n^*, \cdot_n)$ precisely has a primitive root when $n \in \{2, 4, p, 2 \cdot p^e\}$, with p being an odd prime and $e \in \mathbb{Z}^+$.*

Existence of a primitive root

How can primitive roots be found efficiently?

This result, the proof of which can be looked up in [NZ80], gives a necessary and adequate criterion for the existence of a primitive root for group $(\mathbb{Z}_n^*, \cdot_n)$. The question that still remains is how to find such primitive roots efficiently without first trying out all possibilities.

There is no way to give a general response to this question. However, it is possible to construct a group $(\mathbb{Z}_n^*, \cdot_n)$ using a clever choice of n so that a primitive root can be calculated at the same time. Before an appropriate scheme is specified, we first have to highlight some background information and familiarise ourselves with other results.

Theorem 4.7 *If $(S, \circ)$ is a group and $b \in S$, then $(S', \circ)$ with $S' := \{b^a \mid a \in \mathbb{Z}^+\}$ is also a group.*

The proof of this theorem can be read in [CLR90, Section 33.3]. As $S' \subseteq S$, $(S', \circ)$ is also called a *subgroup* of $(S, \circ)$. If b is a primitive root of $(S, \circ)$, then it holds that $S' = S$.

Order of a group and an element

Definition 4.6 *(Order of a group and an element) Let $(S, \circ)$ be a group, e the neutral element of this group and $b \in S$ an arbitrary element of the group. Also let $c \in \mathbb{Z}^+$ be the smallest number so that $b^c = e$, if such a c exists, and if not, let $c := \infty$. Then $|S|$ is called the* **order of the group** $(S, \circ)$ *and c is called the* **order of the element** b.

A relationship exists between the order of a group and that of one of its subgroups, as recorded in the following theorem:

Theorem 4.8 *(LaGrange) If $(G, \circ)$ is a finite group and $(H, \circ)$ is a subgroup of $(G, \circ)$, then the order of H divides the order of G.*

What particularly applies to all elements $b \in G$ is that the order of b divides the order of G. Another statement about the number of elements of a certain order can be made in respect to finite cyclic groups and is gives in the following theorem:

Theorem 4.9 *If $(G, \circ)$ is a finite cyclic group of the order n and d is a divisor of n, then G has exactly $\Phi(d)$ elements of the order d. In particular, G therefore has $\Phi(n)$ elements of the order n (primitive roots).*

Construction of cyclic groups and choice of primitive root

The Theorems 4.6, 4.8 and 4.9 form the basis for the following algorithm that can be used to construct a cyclic group and a primitive root for this group:

❏ Choose a large prime number q so that $p := 2 \cdot q + 1$ is also a prime number.

❏ As p is a prime number, it follows with Theorem 4.6 that $(\mathbb{Z}_p^*, \cdot_p)$ is a cyclic group.

❏ The order of $\mathbb{Z}_p^*$ is $2 \cdot q$ and it holds that $\Phi(2 \cdot q) = \Phi(2) \cdot \Phi(q) = q - 1$ as q is a prime number.

❏ Therefore, according to Theorem 4.9, the probability of selecting a primitive root by randomly choosing an element of $\mathbb{Z}_p^*$ is: $(q - 1)/(2 \cdot q) \approx 1/2$.

❏ An efficient test can be executed with Theorem 4.8 to determine whether a randomly chosen g is a primitive root, because the only thing that needs to be checked is whether $g^2 \equiv 1 \mod p$ or $g^q \equiv 1 \mod p$. If not, then the order has to be of $g = |\mathbb{Z}_p^*|$ since the order of g must be a divisor of the order of $\mathbb{Z}_p^*$ according to Theorem 4.8.

These results give us the mathematical background we need to deal with the problem of the discrete logarithm and its cryptographic applications.

Definition 4.7 *(Discrete logarithm) Let p be a prime number, g a primitive root of $(\mathbb{Z}_p^*, \cdot_p)$ and $c \in \mathbb{Z}_p^*$ any element. Then a number $z \in \mathbb{Z}^+$ exists so that $g^z \equiv c \mod p$. The number z is called the* **discrete logarithm of c modulo p to the basis g.**

Discrete logarithm

For example, 6 is the discrete logarithm of 1 modulo 7 to the basis 3 since $3^6 \equiv 1 \mod 7$.

Calculating the discrete logarithm z for given numbers g, c and p is a computationally difficult problem and the asymptotic runtime of the best known algorithms is exponential to the bit length of prime number p.

Computation of discrete logarithms requires exponential runtime

The problem consequently presents a promising candidate for the design of asymmetric cryptographic schemes. Actually, a number of algorithms have already been proposed and the two best known ones are discussed in the following sections.

4.5 The Diffie–Hellman Key Exchange Algorithm

The Diffie–Hellman key exchange algorithm was published for the first time in May 1976 [DH76]. The paper also introduced the general basic principles of asymmetric cryptography, thus giving fundamental significance to the development of this discipline. Even though the algorithm is not essentially an asymmetric cryptographic scheme, it is nevertheless an important algorithm that is currently being used in numerous cryptographic protocols, not least because it is the only secure scheme of its specific type.

Agreeing a secret over an authentic but not confidential channel

In its original form, the algorithm enables two entities A ('Alice') and B ('Bob') to agree a shared secret over a public channel. In this context a public channel is a communication medium that is not secure from eavesdropping. This means that an attacker E ('Eve,' based on the word 'eavesdrop') can eavesdrop on all messages exchanged between A and B.

It is important for A and B to have the assurance that messages sent through the communication channel cannot be modified because of the threat of 'man-in-the-middle' attacks (more on this below).

The mathematical foundation of the algorithm is the problem of the discrete logarithm in finite cyclic groups introduced in the last section. If A and B want to agree on a shared secret and a public channel is the only communication medium available to them, then they can proceed as follows:

Functioning of Diffie–Hellman algorithm

❑ A chooses a prime number p, a primitive root g of $\mathbb{Z}_p^*$ and a random number $q \in \mathbb{Z}_p^*$. A and B can either agree on values for p and g before the first protocol run or A can send these values with her first message to B.

A now computes $v = g^q$ MOD p and sends to B: (p, g, v)

❑ B also selects a random number r, computes $w = g^r$ MOD p and sends w to A.

❑ A computes $s = w^q$ MOD p.

❑ B computes $s' = v^r$ MOD p.

Since $g^{(q \cdot r)}$ MOD $p = g^{(r \cdot q)}$ MOD p, it holds $s = s'$ so that A and B have agreed on the same number. An attacker E who is eavesdropping on the public channel can only figure out the number s computed by A and B if she can calculate the discrete logarithm modulo p to the basis g from one of the two numbers v or w. The prime number p chosen by A and B need, therefore, only be large enough so that an attacker cannot discover their shared secret — dependent on the runtime, exponential in the bit length of p, of the computation of the discrete logarithm. ❑

In practice it is not mandatory that g be a primitive root of $(\mathbb{Z}_p^*, \cdot_p)$, and it suffices if g creates a sufficiently large subgroup of $(\mathbb{Z}_p^*, \cdot_p)$.

If a potential attacker E is able to modify messages on the path between A and B, she can execute what is called a *man-in-the-middle attack*:

Man-in-the-middle attack

❏ E randomly selects two numbers q', r' and computes:

$$v' = g^{q'} \text{ MOD } p \quad \text{and} \quad w' = g^{r'} \text{ MOD } p$$

❏ The message (p, g, v) sent by A is intercepted by E who instead sends message (p, g, v') to B.

❏ E also intercepts message (w) sent by B and instead sends (w') to A.

❏ A computation of the supposed shared secret now produces the following situation:

  ❏ A computes $s_1 = w'^q \text{ MOD } p = v^{r'} \text{ MOD } p$, with the last computation being one executed by E.

  ❏ B computes $s_2 = v'^r \text{ MOD } p = w^{q'} \text{ MOD } p$, with the last computation again being one executed by E.

In essence, A and E have agreed on a shared secret s_1, but E and B have agreed on a different secret s_2.

If A and B use their respective supposed secret to encrypt messages they exchange over a public channel, E can intercept these messages, decrypt them and then send them newly encrypted to the actual receiver. From this point onwards E actually has to manipulate all message traffic secured through a 'shared secret' between A and B if he or she wants to avoid discovery.

An attacker can subsequently control an entire communication

This example demonstrates the importance of being 'certain' about a person with whom one has negotiated a key, i.e., the authenticity of key-exchange messages is vital. We will take another look at this aspect in Chapter 7.

Authenticity of key-exchange messages is vital!

4.6 The ElGamal Algorithm

The ElGamal algorithm was published in 1985 [ElG85]. The algorithm can be used either for digital signatures or for encrypting data although, unlike RSA, the two computation variants differ from one another. Like the Diffie–Hellman key-exchange scheme, it is based on the problem of the discrete logarithm. However, as it uses additive as well as multiplicative operations, it is computed in finite fields rather than in finite groups like Diffie–Hellman. (However, the group $(\mathbb{Z}_p^*, \cdot_p)$ normally used for Diffie–Hellman meets the requirements of a field if 0 and the operation $+_p$ are added, so in practice there is actually no difference between the two mathematical structures).

Calculating a key pair   The following steps are necessary to calculate a key pair for the ElGamal algorithm:

❑ Choose a large prime number p, a primitive root g of the multiplicative group $(\mathbb{Z}_p^*, \cdot_p)$ and a random number v so that $1 \leq v \leq p - 2$.

❑ Compute: $y = g^v \text{ MOD } p$

❑ The public key is: (y, g, p)

❑ The private key is: (v, g, p)

ElGamal signatures   The following scheme is used to sign a message $m < p$:

❑ Choose a large random number k so that k is relatively prime to $p - 1$.

❑ Compute: $r = g^k \text{ MOD } p$

❑ Use the extended Euclidean algorithm k^{-1} to calculate the inverse element to $k \mod (p - 1)$.

❑ Compute: $s = k^{-1} \cdot (m - v \cdot r) \text{ MOD } (p - 1)$

❑ The signature of the message m is (r, s).

Signature check   Checking whether $y^r \cdot r^s \equiv g^m \mod p$ is all that is needed to verify a signature. The following lemma is helpful for proofing the correctness of this scheme:

Lemma 4.3 *Let p be a prime number and g a generator of $(\mathbb{Z}_p^*, \cdot_p)$, it then holds that:*

$$i \equiv j \mod (p - 1) \Rightarrow g^i \equiv g^j \mod p$$

For proof of this lemma one has to be sure that

❑ $i \equiv j \mod (p - 1) \Rightarrow \exists k \in \mathbb{Z}^+ : (i - j) = (p - 1) \cdot k$

❏ Therefore, according to Theorem 4.4 (Euclidean):

$$g^{(i-j)} \equiv g^{(p-1) \cdot k} \equiv 1^k \equiv 1 \mod p$$
$$\Rightarrow g^i \equiv g^j \mod p \quad \square$$

The following calculation proves the correctness of the ElGamal signature check:

Correctness of ElGamal signature check

$$
\begin{aligned}
& s \equiv k^{-1} \cdot (m - v \cdot r) && \mod (p-1) \\
\Leftrightarrow \quad & k \cdot s \equiv m - v \cdot r && \mod (p-1) \\
\Leftrightarrow \quad & m \equiv v \cdot r + k \cdot s && \mod (p-1) \\
\Rightarrow \quad & g^m \equiv g^{(v \cdot r + k \cdot s)} && \mod p && \textbf{(Lemma 4.3)} \\
\Leftrightarrow \quad & g^m \equiv g^{(v \cdot r)} \cdot g^{(k \cdot s)} && \mod p \\
\Leftrightarrow \quad & g^m \equiv y^r \cdot r^s && \mod p \quad \square
\end{aligned}
$$

The security of ElGamal signatures is based on the fact that an attacker needs the private key v to compute value s. Therefore, the attacker must compute the discrete logarithm of y modulo p to the basis g in order to forge signatures.

Security of ElGamal signatures

It is vital for the security of ElGamal signatures that a new random number k is chosen for each signature. Otherwise an attacker will be able to compute v from two different messages and their signatures, formed using the same k [MOV97, Note 11.66.ii].

With the ElGamal scheme messages should not be signed directly and instead a *cryptographic hash value* (see also Chapter 5) should be computed that is then signed using an ElGamal algorithm. If messages are signed directly with an ElGamal algorithm, an attacker may be able to construct a message with a valid signature in certain circumstances [MOV97, Note 11.66.iii].

With the ElGamal scheme, encryption of a message m with $m < p$ is realised as follows:

ElGamal encryption

❏ Choose a random number $k \in \mathbb{Z}^+$ with $k \leq p - 1$.

❏ Compute: $r = g^k \bmod p$

❏ Compute: $s = m \cdot y^k \bmod p$

❏ The ciphertext is (r, s). Note that the ciphertext is twice as long as the original message m.

The following two steps are needed to decrypt a message:

ElGamal decryption

❏ The private key v is first used to compute $r^{(p-1-v)} \bmod p = r^{-v} \bmod p$ (again Theorem 4.4 is adopted).

❏ r^{-v} can then be used to compute m, since:

$$m = r^{-v} \cdot s \text{ MOD } p$$

Correctness of ElGamal decryption

The following congruence is used to prove the correctness of ElGamal decryption:

$$r^{-v} \cdot s \equiv r^{-v} \cdot m \cdot y^k$$
$$\equiv g^{(-v \cdot k)} \cdot m \cdot y^k$$
$$\equiv g^{(-v \cdot k)} \cdot m \cdot g^{(v \cdot k)}$$
$$\equiv m \qquad\qquad \text{mod } p \quad \square$$

Security of ElGamal encryption

The security of ElGamal encryption is based on the fact that an attacker must know the discrete algorithm v of number y modulo p to the basis g in order to decipher an encrypted message.

As with ElGamal signatures, it is vital for the security of ElGamal encryption that a new random number k is chosen for each message because an attacker is otherwise able to compute v [MOV97, Note 8.23.ii].

4.7   Security of Asymmetric Cryptographic Schemes

Generalisation to other mathematical groups and fields

The algorithms presented in this chapter were invented for the multiplicative groups $(\mathbb{Z}_p^*, \cdot_p)$ and the fields $(\mathbb{Z}_p, +_p, \cdot_p)$. During the 1980s it was shown that these algorithms could be generalised and also used with other groups or fields.

The main motivation for this generalisation lies in the security sought for these algorithms, which until now has been directly related to key length.

Success in the field of prime number tests and factorisation

Since the invention of the RSA algorithm there has been considerable success in mathematical research in the fields of prime number tests, prime factor analysis and the computation of the discrete logarithm. Not least this success can be attributed to the importance of these results for the security of the algorithms based on these problems:

❏ When the *RSA-129 challenge*, the factoristion of a 129-digit (≈ 428 bit) long number, was published in 1977, it was estimated that it would take around 40 quadrillion years to solve the problem. In actual fact, it took less than twenty years;

in 1994 it took only eight months for a network of computers communicating over the Internet with a computing time equivalent to around 5000 MIPS-years[1] to solve the task.

However, the reason why the original estimate was refuted was not because of a false calculation in 1977 but because the estimate was based on other assumptions. The estimate may have been totally correct in respect of the algorithms for prime factor factorisation known in 1977, but what happened is that advances in mathematics in the following 17 years enabled the same tasks to be completed with considerably less processing time.

❑ Then only two years later, in 1996, a new algorithm was used successfully in factorising a 130-digit ($\approx$ 431 bit) long number with a computational effort of about 500 MIPS-years. A problem that was eight-times larger was thus solved with a tenth of the computational effort.

These examples show the risks to the security of asymmetric cryptographic algorithms that result from mathematical progress. As a consequence, the minimal requirements for key length have to be corrected upward at regular intervals. They are currently between about 1024 and 2048 bits.

Risks due to mathematical progress

The longer key lengths indeed increase the level of security but, on the downside, they result in a reduction in processing speed. The reason is the required exponentiation of individual key values during the execution of the algorithms, and this operation involving such long keys is relatively inefficient.

Performance loss due to increased key length

The most efficient factorisation algorithms are based on specific properties of $(\mathbb{Z}_p^*, \cdot_p)$ and $(\mathbb{Z}_p, +_p, \cdot_p)$. Therefore, a number of mathematicians have tried to generalise these algorithms to other mathematical structures that are not as receptive to optimised factorisation algorithms or to algorithms used to compute the discrete logarithm.

The idea is the acceptance of more complicated basic operations $(\oplus, \odot)$ on the elements of a group or a field in return for the guarantee that a shorter key length will provide a comparable level of security as a longer key in $(\mathbb{Z}_p^*, \cdot_p)$ or $(\mathbb{Z}_p, +_p, \cdot_p)$. This re-

---

[1] One MIPS-year corresponds to the processing time that a computer capable of executing one million instructions per second (*million instructions per second, MIPS*), can provide in one year. A Pentium-200 processor has a computing capacity of 50 MIPS.

sults in a reduction of the computational effort algorithms require overall while a certain level of security is maintained.

Cryptography with elliptic curves

The mathematical structure that has proven to be very promising for this purpose is the *group of points on an elliptic curve over a finite field*. The mathematical operations on this field can be implemented efficiently in hardware as well as in software. The best algorithms for factorisation or computation of the discrete logarithm in this field are by far less efficient that those for the field $(\mathbb{Z}_p, +_p, \cdot_p)$. Although Menezes has found sub-exponential algorithms for the computation of the discrete logarithm for the special case of what is called *super-singular elliptic curves*, these solutions are not appropriate for use in general cases [Men93].

A comprehensible introduction into cryptography with elliptic curves would extend beyond the bounds of what is appropriate as mathematical background in this book. We therefore refer the reader to other literature available on this topic [Kob87, Men93].

4.8 Summary

Basic idea of asymmetric cryptography

Asymmetric cryptography enables two matching but distinct keys to be used for encryption and decryption or for signing and signature checking.

The algorithms that are still considered secure today and have also proven effective in practice are:

❏ *RSA algorithm*, based on the difficulty of factorising large integers

❏ *Diffie–Hellman key exchange* (not actually an asymmetric scheme), based on the difficulty of calculating the discrete logarithm of the number y modulo of a prime number p to a given basis g in finite fields

❏ *ElGamal algorithm*, which is also based on the problem of the discrete logarithm

Security of asymmetric algorithms

Because the security of these algorithms relies on the complexity of the mathematical problems named, advances in the field of mathematical algorithms represent the greatest threat to these schemes. One should always bear in mind that, unlike forecasting technological progress in the field of processor construction, estimating the impact of algorithmic advances is very difficult, if not impossible.

The following practical considerations in particular have to be taken into account with the use of asymmetric cryptographic algorithms:

Practical considerations

❏ Asymmetric cryptographic operations are magnitudes (around the factor or 100 to 1000) slower than symmetric schemes.

❏ Consequently, they are seldom used for encrypting or signing bulk data. Instead symmetric encryption schemes or cryptographic hash functions (see also Chapter 5) are used for this purpose, and asymmetric schemes are reserved for the encryption of so called session keys or for the signing of cryptographic hash values.

4.9 Supplemental Reading

[Bre89] offers a good introduction to the mathematical background of prime number tests and factorisation, whereas [NZ80, Kob87] provides a general introduction to number theory, with the latter book also highlighting cryptographic applications. Cryptography with elliptic curves is discussed in detail in [Men93]. An easy-to-understand introduction into the subject with emphasis on the implementation and computing time of algorithms can also be found in [CLR90, Chapter 33]. The subject is also covered in detail in [MOV97]. We also highly recommend a study of the three main original articles [DH76, RSA78, ElG85].

[Bre89]   BRESSOUD, D. M.: *Factorization and Primality Testing*. Springer, 1989

[CLR90]   CORMEN, T. H.; LEISERSON, C. E. ; RIVEST, R. L.: *Introduction to Algorithms*. MIT Press, 1990

[DH76]   DIFFIE, W.; HELLMAN, M. E.: New Directions in Cryptography. In: *Trans. IEEE Inform. Theory, IT-22*, 1976, pp. 644–654

[ElG85]   ELGAMAL, T.: A Public Key Cryptosystem and a Signature Scheme Based on Discrete Logarithms. In: *IEEE Transactions on Information Theory* 31, July 1985, No. 4, pp. 469–472

[Kob87]   KOBLITZ, N.: *A Course in Number Theory and Cryptography*. Springer, 1987

[Men93]   MENEZES, A. J.: *Elliptic Curve Public Key Cryptosystems*. Kluwer Academic Publishers, 1993

[MOV97]   MENEZES, A.; VAN OORSCHOT, P.; VANSTONE, S.: *Handbook of Applied Cryptography*. CRC Press LLC, 1997

[NZ80]   NIVEN, I.; ZUCKERMAN, H.: *An Introduction to the Theory of Numbers*. 4th edition, John Wiley & Sons, 1980

[RSA78]   RIVEST, R.; SHAMIR, A.; ADLEMAN, L.: A Method for Obtaining Digital Signatures and Public Key Cryptosystems. In: *Communications of the ACM*, February 1978

4.10   Exercises

1.   Which properties must a practical asymmetric cryptographic scheme have?

2.   Why is it important for the sender of an asymmetrically encrypted message to check the authenticity of the public key of the receiver?

3.   Using the extended Euclidean algorithm, compute the greatest common divisor of 210 and 126 as well as the two coefficients of the corresponding linear combination.

4.   Using Theorem 4.2, show that $\sqrt{2}$ is not a rational number.

5.   Compute $\Phi(21)$.

6.   Explain how the RSA algorithm functions.

7.   The following requirement of the RSA key appears in the 1988 version of International Standard X.509: '*It must be ensured that $e > log_2(n)$ to prevent attack by taking the e-th root modulo n to disclose the plaintext.*'

     Although the condition required is correct, the reasoning given is not right. Explain the error in the reasoning and give the right reasoning for this requirement.

8.   Find the primitive root of $(\mathbb{Z}_{11}^*, \cdot_{11})$.

9. Compute the discrete logarithm of 1 modulo 5 to basis 3 and of 1 modulo 11 to basis 3.

10. How does a man-in-the-middle attack work on Diffie–Hellman key exchange and how can it be prevented or detected?

11. Is a primitive root of group $(\mathbb{Z}_p^*, \cdot_p)$ really needed with the ElGamal algorithm or is it enough if g produces a sufficiently large subgroup of $(\mathbb{Z}_p^*, \cdot_p)$?

12. What is the practical disadvantage of transmitting messages encrypted with the ElGamal algorithm compared to RSA-encrypted messages?

5 Cryptographic Check Values

This chapter looks at how a check value can be added to data so that the data can be checked at a later point in time to determine whether it has been accidentally or intentionally modified. In the first section, the *cryptographic check values* developed for this purpose are delineated from *error checksums* familiar from data communications. The requirements of these check values are discussed and the two principle categories of cryptographic check values, *modification detection codes* and *message authentication codes*, are introduced. The two sections that follow discuss both categories in detail and Section 5.4 concludes with a description of hybrid schemes.

5.1 Requirements and Classification

In data communications it is common to calculate an error checksum for transmitted data units. This checksum enables a receiver to check whether the message was altered during transmission. The computation of *parity bits*, or the variant *bit-interleaved parity*, and the *Cyclic Redundancy Check (CRC)* are two known schemes for this task [PD00, Section 2.4].

Error checksums

The usefulness of these schemes motivates an interest in computing comparable checksums that enable the checking of data during transmission (or storage, etc.) to determine whether it has been modified. However, one has to take into account that it makes a big difference whether data is altered due to *accidental error* or due to *intentional modification*.

For example, if an attacker wants to modify a protocol data unit that is protected from error by a CRC-checksum, he or she must only ensure that the forged data unit has a valid CRC-checksum. This scheme is no more effective at protecting against intentional modification than other error detection values because an attacker simply has to compute a new CRC-checksum for the modified data unit.

CRC is not suitable as a cryptographic check value

One could of course be naïve and decide to encipher the CRC-checksum of the original message using a previously negotiated session key before transmission to prevent an attacker from simply computing a new encrypted CRC. In fact, this approach was pursued during the standardisation of the wireless local area network standard IEEE 802.11. Unfortunately, it does not offer adequate protection from intentional modification, as we will see in Chapter 15.

This is due to the fact that a cryptographic check value must satisfy a number of requirements that make it impossible for an attacker to forge such a check value and that cannot be provided by an error checksum.

Principle categories   Two principal categories of cryptographic check values can be distinguished at a higher level of consideration:

❏ *Modification Detection Codes (MDC)* that in a certain way implement 'cryptographic fingerprints' of messages.

❏ *Message Authentication Codes (MAC)* that also include a previously negotiated 'secret' i.e., a session key into the computation, thereby enabling a direct verification of the authenticity of a message.

Main application and design goals   The main application of these schemes is in checking the authenticity of a message although the concrete design goals of both categories differ:

❏ An MDC represents the digital fingerprint of a message that can be signed with an asymmetric cryptographic scheme such as RSA or ElGamal. Because it is only the 'fingerprint' that is being signed, it should not be computationally feasible for an attacker to construct two messages that are mapped to the same MDC. This would enable an attacker to use a signed MDC for a different message.

❏ A MAC, on the other hand, allows verification that the creator of a message knows the secret key K that was used in the computation of the MAC of this message, and consequently the message cannot be modified without knowledge of the key.

Other applications   In addition to the main application just described, other applications of cryptographic check values are possible but these must be viewed with a certain caution. They tend to use the schemes for purposes that were not considered in the design process:

❏ verification of knowledge of a particular data value;

❏ generation of sessions keys;

❏ generation of pseudo-random numbers (see also Chapter 6).

Depending on the application, cryptographic check values must sometimes satisfy other requirements. The following section takes a detailed look at both categories of cryptographic check values with the main emphasis on how they function and how they are used to detect modifications.

5.2 Modification Detection Codes

Modification detection codes are computed using *cryptographic hash functions*. These hash functions are distinguished from conventional hash functions:

Definition 5.1 *A* **hash function** *is a function h that has the two following properties:*

❏ **Compression:** *The function h maps input values of an arbitrary finite bit length to output values of a fixed bit length which is specific for h.*

❏ **Ease of computation:** *If h and an input value x are given, then it is easy to compute $h(x)$.*

Hash function

In comparison, cryptographic hash functions must satisfy yet other requirements:

Definition 5.2 *A* **cryptographic hash function** *is a hash function h that additionally has the following properties:*

❏ **Pre-image resistance:** *This property requires that it should not be computationally feasible to calculate an image x for a given value y so that $h(x) = y$.*

❏ **2nd pre-image resistance:** *This property requires that for a given value x it should not be computationally feasible to find a second value x' that is mapped to the same value $h(x) = h(x')$.*

❏ **Collision resistance:** *This property requires that it should not be computationally feasible to calculate two images $x_1 \neq x_2$ that are mapped to the same value $h(x_1) = h(x_2)$.*

Cryptographic hash function

Other potential requirements   Depending on the intended application, cryptographic hash functions may sometimes have to meet further requirements, e.g., *partial pre-image resistance*. This means that if only t bits of the image are not known, an average of 2^{t-1} computing steps are needed to calculate the remaining t bits.

The general structure of common cryptographic hash functions, along with two important representatives of this class of function, will be explained below. The section concludes with a discussion on potential attacks on modification detection codes, thus also providing a justification for the properties listed above.

5.2.1   General Structure of Cryptographic Hash Functions

Comparable to the development of symmetric encryption algorithms that make use of certain established structures, such as the Feistel network, the design of the most commonly used cryptographic hash functions today is based on the underlying general structure shown in Figure 5.1.

Figure 5.1
General structure of cryptographic hash functions

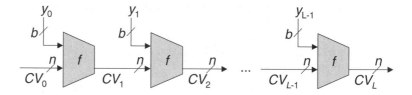

Padding and appending of length field   A length field is appended to an arbitrary message y and the message is padded to an integer multiple of the block size b of the cryptographic hash function. Let $(y_0 \,\|\, y_1 \,\|\, ... \,\|\, y_{L-1})$ be the resulting message[1] that is structured into L blocks of block size b.

Chaining with a chaining value   In order to compute the hash value $h(y_0 \,\|\, y_1 \,\|\, ... \,\|\, y_{L-1})$, the message is processed block by block, and in each step a so-called *chaining value* CV_{i+1} of length n bits is computed from the message block y_i and the chaining value CV_i using a *compression function* f that maps $(n + b)$ bits to n bits: $CV_{i+1} := f(CV_i, y_i)$

Initialisation vector   The first chaining value CV_0 requires an initialisation vector IV that is normally defined by the cryptographic hash function. The cryptographic hash value of the message is then the value of the chaining value after the last iteration step: $h(y_0 \,\|\, y_1 \,\|\, ... \,\|\, y_{L-1}) := CV_L$.

---

[1]In this context '$\|$' denotes the concatenation of blocks.

The functioning of such a cryptographic hash function can be summarised as follows:

$$CV_0 := IV$$
$$CV_i := f(CV_{i-1}, y_{i-1}) \qquad 1 \leq i \leq L$$
$$H(y) := CV_L$$

It has been proved that the collision resistance of compression function f for this structure also ensures the collision resistance of the resulting cryptographic hash function h [Mer89].

Consequently, the analysis of cryptographic hash functions normally concentrates on the internal structure of the compression function f and tries to find efficient techniques for creating collisions for a single execution of function f.

The cryptanalysis of cryptographic hash functions concentrates on the compression function

The minimum bit length required by hash value $h(x)$ today is normally 160 bits. This requirement was established due to the threat of so called *birthday attacks* which will be discussed later (see also Section 5.2.4). With a block length of 160 bits, such an attack requires an average effort of the magnitude of 2^{80}, which according to today's knowledge is considered impractical.

The two currently most popular cryptographic hash functions *MD5* and *SHA-1*, which both follow the general structure just outlined, are explained in the following two sections.

5.2.2 MD5

The cryptographic hash function *Message Digest 5 (MD5)* [Riv92] was designed by Rivest as a successor to the function *Message Digest 4 (MD4)* [Riv91].

MD5 is constructed according to the structure described above. First one '1'-bit and so many '0'-bits are added to message y that the length of the resulting message is congruent to 448 modulo 512. Then the length of the original message is appended as a 64-bit long length field so that the length of the resulting message is an integer multiple of 512. This message is then divided into blocks y_i of length $b = 512$ bits.

Padding and addition of length field

The length of the chaining value is $n = 128$ bits and the value is structured into four 32-bit long registers A, B, C, D. The following preset initialisation vectors are written into the register for the initialisation:

Chaining value and initialisation vector

$$A := 0x\ 01\ 23\ 45\ 67 \qquad B := 0x\ 89\ AB\ CD\ EF$$
$$C := 0x\ FE\ DC\ BA\ 98 \qquad D := 0x\ 76\ 54\ 32\ 10$$

This initialisation vector is given in a Little-Endian format, i.e., the lower-valued bits are noted at the end.

Compression function   Each block y_i of the message is then mixed with chaining value CV_i through function f, which is realised internally through four rounds, each with 16 steps. Each round is based on a similar structure and uses a table containing 64 constant values each 32 bits long. In addition, a round-specific logical function g is used in each round. Figure 5.2 shows a block diagram of an individual step.

Figure 5.2
An iteration of the
MD5 function

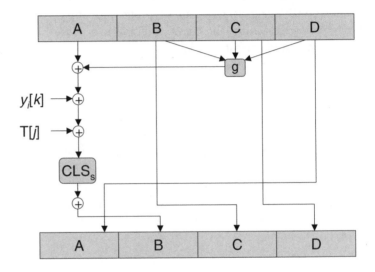

Iteration step process   In each step the contents of registers B, C and D are linked to the round-specific function g and the function value is XORed (XOR, represented by the character $\oplus$) to the contents of register A. The resulting value is then XORed to $y_i[k]$, the kth 32-bit word of the ith message block, and than XORed again with $T[j]$, the jth table entry, with j incremented modulo 64 with each step. The value that results is then cyclically left shifted by s bits (Cyclical Left Shift, CLS) before it is written into register B for the next step. The quantity s by which each shift takes place is fixed for each round (the exact specification will not be dealt with here). Furthermore, the contents of register D are written into A, those of register B into C and those of register C into D.

The MD5 value of message y is then the contents of the registers after the last data block y_i has been processed.

Security of MD5   Concerning MD5's security, it should be noted that each bit of the 128-bit long output depends on each bit of input message y.

However, the length of the output with 128 bits is generally still judged to be too short (also see Section 5.2.4).

Important advances were made in the analysis of MD5 between 1992 and 1996. In 1996 Dobbertin published an attack that produced a collison for the function f of MD5. Even if this attack was not expanded to a collision for the full MD5 function, it created considerable concern for the security of MD5. The company RSA Laboratories, which holds the rights to MD5, reacted to the announcement with the following comments: *'Existing signatures formed using MD5 are not at risk and, while MD5 is still suitable for a variety of applications (particularly those that solely rely on the one-way property of MD5 and on the 'randomness' of its output), it should not be used in future for applications that require the property of collision resistance.'* [Rob96]

Cryptanalysis of MD5

MD5 should not be used for applications that require collision resistance!

5.2.3   SHA-1

The function *Secure Hash Algorithm 1 (SHA-1)* was developed by the National Security Agency (NSA). The NSA designed SHA-1 according to the structure of the MD4 function, so that both functions follow the general structure described above.

SHA-1 works on block lengths of 512 bits and produces an output value of length 160 bits. Its procedure for preparing a message y and separating it into blocks y_i is identical to that of MD5.

Padding and addition of length field

The chaining value of SHA-1 is the 32-bit longer output divided into five registers of length 32 bits. For initialisation these registers are filled with the following constants:

Chaining value and initialisation vector

$$A := 0x\,67\,45\,23\,01 \qquad B := 0x\,EF\,CD\,AB\,89$$
$$C := 0x\,98\,BA\,DC\,FE \qquad D := 0x\,10\,32\,54\,76$$
$$E := 0x\,C3\,D2\,E1\,F0$$

For the first four registers this value corresponds to that of MD5. However, in contrast to MD5, the definition of SHA-1 is given for a Big-Endian architecture in which higher-valued octets are stored according to lower-valued ones so that the construction initially looks different.

Each message block y_i is mixed with the content of chaining value CV_i through application of function f resulting in the new chaining value CV_{i+1}. The function f is implemented in four rounds with 20 steps each. The rounds all have the same structure but a specific logical function f_1, f_2, f_3 or f_4 is used per round. In

Compression function

5 Cryptographic Check Values

addition, each step makes use of a fixed additive constant K_t that remains unchanged during a round.

Figure 5.3

An iteration of the SHA-1 function

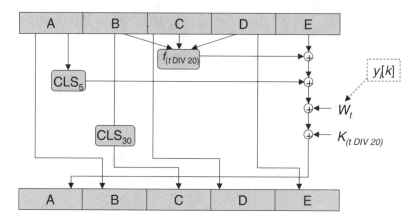

Figure 5.3 shows a block diagram of the internal structure of a step. The current message block y_i is logically integrated into the processing through value W_t, which is defined according to the following recurrence relationship:

$$t \in \{0, ..., 15\} \quad \Rightarrow W_t := y_i[t]$$
$$t \in \{16, ..., 79\} \quad \Rightarrow W_t := CLS_1(W_{t-16} \oplus W_{t-14} \oplus W_{t-8} \oplus W_{t-3})$$

After the 79th step each one of registers A, B, C, D, E is added modulo 2^{32} to the content of the corresponding register before the 0th step, and the result is written into the relevant register for processing the next message block.

As with MD5, the SHA-1 value of message y is the content of the register after the last data block y_i has been processed.

Security of SHA-1   In terms of security, it should be noted that SHA-1 offers better protection against brute-force and birthday attacks (see below) than does MD5 because of its longer output value. There has been no publicity yet about any considerable cryptanalytic success against compression function f of SHA-1. However, it should be mentioned that, unlike MD5, the design criteria for SHA-1 were not published and, as a consequence, cryptanalysis of SHA-1 is more difficult.

Comparison with MD5   A further comparison shows that SHA-1 is around 25% slower in processing than MD5, which is a direct consequence of the processing effort required for the 25% longer chaining value. Both algorithms are simple to define and implement and do not require any extensive programmes or substitution tables. Concerning the

Little-Endian or Big-Endian architecture (selected as indicated above), there is no particular advantage to choosing one function over the other, and the presumption is that the choice depended more on the computer platforms that were available to developers of MD5 and SHA-1 than on an actual design decision. RSA Laboratories themselves recommend SHA-1 or the function *RipeMD-160* (not discussed here) for applications that rely on collision resistance [Rob96].

SHA-1 is suitable for applications that require collision resistance

5.2.4   Attacks on Modification Detection Codes

A modification detection code should represent a tamper-proof 'digital fingerprint' of a message and allow it to be signed using an asymmetric cryptographic scheme to verify the authenticity of the message. It is vital that it is not computationally feasible for two messages $x_1 \neq x_2$ to be created with the same modification detection code $MDC(x_1) = MDC(x_2)$. Otherwise, an attacker E could try to recover a digital signature for a 'harmless' message x_1 from an entity A and use it for a 'not so harmless' message x_2 (for example, an electronic order for goods or services).

The collision resistance of the cryptographic hash function used for digital signatures is therefore particularly important. This aspect will be examined closely below.

For this purpose we will first take a look at what is called the *birthday phenomenon*, which is directly responsible for the minimum length requirement for the output of cryptographic hash functions.

Birthday phenomenon

In its basic form the birthday phenomenon supplies an answer to the question of 'how many people have to be in a room so that the probability of two people having the same birthday is at least 50%'. For reasons of simple computation, we leave leap years out of the equation and assume that each birthday has the same probability.

First we define $P(n, k)$ as the probability that at least one duplication will occur with k randomly selected variables each capable of assuming an equally probable value between 1 and n. This corresponds to the random experiment of having a box of n different balls and pulling out a ball k times with the respective ball being put back in the box each time.

$P(n, k) \approx$ at least one duplication

We also define $Q(n, k)$ as the probability that no duplication will occur with the experiment described. If a ball should be pulled out more than once, we can therefore select the first ball from n possible balls, the second ball from $n-1$ possible balls and so forth. Thus the number of different possibilities of pulling out k balls

$Q(n, k) \approx$ no duplication

from n different balls without duplication is on the order of:

$$N = n \cdot (n-1) \cdot ... \cdot (n-k+1) = \frac{n!}{(n-k)!}$$

The number of different possibilities of pulling out k balls from n possible balls and returning them to the box is n^k. Therefore:

$$Q(n,k) = \frac{N}{n^k} = \frac{n!}{(n-k)! \cdot n^k}$$

Computing $P(n,k)$ So it holds that:

$$P(n,k) = 1 - Q(n,k)$$
$$= 1 - \frac{n!}{(n-k)! \cdot n^k}$$
$$= 1 - \frac{n \cdot (n-1) \cdot ... \cdot (n-k+1)}{n^k}$$
$$= 1 - \left[\frac{n-1}{n} \cdot \frac{n-2}{n} \cdot ... \frac{n-k+1}{n} \right]$$
$$= 1 - \left[\left(1 - \frac{1}{n}\right) \cdot \left(1 - \frac{2}{n}\right) \cdot ... \cdot \left(1 - \frac{k-1}{n}\right) \right]$$

Using the inequation $\forall\, x \geq 0 \colon\ (1-x) \leq e^{-x}$ we then have:

$$P(n,k) > 1 - \left[\left(e^{\frac{-1}{n}}\right) \cdot \left(e^{\frac{-2}{n}}\right) \cdot ... \cdot \left(e^{\frac{-(k-1)}{n}}\right) \right]$$
$$= 1 - e^{-\left[\frac{1}{n} + \frac{2}{n} + ... + \frac{k-1}{n}\right]}$$
$$= 1 - e^{-\frac{k \cdot (k-1)}{2 \cdot n}}$$

The following equation was used in the last step:

$$1 + 2 + ... + (k-1) = \frac{(k^2 - k)}{2}$$

Returning to the original question of how many people would have to be together in a room so that the probability that at least two of these people have the same birthday is greater than or the equal to 0.5, we have to solve the following equation:

$$\frac{1}{2} = 1 - e^{\frac{-k \cdot (k-1)}{2 \cdot n}}$$
$$\Leftrightarrow \qquad 2 = e^{\frac{k \cdot (k-1)}{2 \cdot n}}$$
$$\Leftrightarrow \qquad ln(2) = \frac{k \cdot (k-1)}{2 \cdot n}$$

Estimating for large k   For large k we can estimate $k \cdot (k-1)$ through k^2 and arrive at:

$$k \approx \sqrt{2 \cdot \ln(2) \cdot n} \approx 1.18 \cdot \sqrt{n}$$

For $n = 365$ the result with $k \approx 22.54$ is quite close to the actual value $k = 23$.

We have therefore shown that the quantity k of the values that have to be selected from n possible values in order to arrive at two identical values with a probability greater than or equal to 0.5 is on the magnitude of $\sqrt{n}$.

This fact can be exploited for attacks on modification detection codes [Yuv79]:

Use for attacks

❏ Assume that E wants to induce A to sign a message m_1 that A normally would not sign (e.g. an electronic order). E knows that A signs her message by first computing a modification detection code $MDC(m)$ of length r bits and then signs it with her private key.

❏ Because E also knows that A would not sign message m_1, she will not present it to her directly for signing. If she tried to construct a second harmless message m_2, which is mapped to the same modification detection code $MDC(m_2) = MDC(m_1)$, the average computational effort required would be of the magnitude of 2^{r-1} because on average she would have to search through half the search space.

❏ Instead E takes advantage of the birthday phenomenon and, using the two messages m_1 and m_2 as a basis, starts to construct variations of these messages m_1' and m_2' until she finds a combination that is mapped to the same modification detection code $MDC(m_1') = MDC(m_2')$. For example, E can use character combinations in the form of '<space>, <backspace>' or replace individual words with synonyms to create messages that are semantically identical but have different coding.

❏ As we know from the birthday phenomenon, on average E requires $\sqrt{2^r} = 2^{r/2}$ variations of both messages to ensure that the probability of a successful collision with $MDC(m_1') = MDC(m_2')$ is greater than or equal to 0.5.

❏ Because E has to create and store the variations of the messages, the storage and computation effort is of the magnitude of $2^{r/2}$.

❏ After E has found two suitable messages m_1' and m_2', she presents message m_2' to A for signing. The signature generated by Alice likewise represents a valid signature of message m_1' so that E can subsequently make the claim that A signed message m_1'.

Birthday attacks Attacks that follow this pattern are called *birthday attacks*. According to current estimates, a cryptographic hash function should have an output value of a 160-bit minimum length to make it secure from this attack technique as an average effort of 2^{80} is not considered feasible.

Let us assume that A uses a cryptographic hash function that produces a 96-bit length output and then signs it with a 2048-bit long RSA key. Because it is not feasible to break a 2048-bit long RSA key with the algorithms currently known, one would assume that the security a signature with such a key offers cannot be compromised. However, adequate security is not provided due to the possibility of a birthday attack the average effort of which in this case is on the magnitude of 2^{48}. This example shows the importance of evaluating all known attack techniques on separate cryptographic algorithms used as well as on task-specific combinations when assessing the security of a cryptographic solution. One can make an analogy to a chain that is known only to be as strong as its weakest link.

With cryptographic 'solutions' it is always necessary to evaluate concrete combinations of separate schemes

5.3   Message Authentication Codes

Definition 5.3 *An* **algorithm for computing message authentication codes** *is a family of functions h_K that is parameterised through a secret key K and has the following properties:*

❏ **Compression:** *The functions h_K map input values of an arbitrarily finite bit length to an output value of a bit length fixed for h_K. This is also called a* **Message Authentication Code (MAC).**

❏ **Simplicity of computation:** *If a family of functions h_K, the key K and an input value x exist, then $h_K(x)$ can be computed with little effort.*

❏ **Computation resistance of MACs to new messages:** *This property means that a valid MAC $h_K(x)$ cannot be computed from a range of known pairs $(x_i, h_K(x_i))$ for a new message $x \neq x_i \, \forall \, i$ if the key K is not known.*

At this juncture it should be mentioned that the latter property, computation resistance of message authentication codes, implies that it should not be possible to calculate key K from a range of pairs $(x_i, h_K(x_i))$ ('key non-recovery'). However, a reverse conclusion cannot be reached because, depending on the structure of the MAC algorithm, knowledge of key K is not required to forge a valid MAC for a message from a range of known pairs of messages and corresponding MACs.

Calculation of key K not always necessary for forging a MAC

We will use the following definition of an insecure message authentication scheme to present an instructional example:

Example of an insecure MAC

❏ Input: Message $m = (x_1, x_2, ..., x_n)$ in which each x_i represents a 64-bit long message block and key K;

❏ Compute: $\Delta(m) := x_1 \oplus x_2 \oplus ... \oplus x_n$, with '$\oplus$' denoting the bit-wise XOR operation;

❏ Ouput: MAC $C_K(m) := E(K, \Delta(m))$, with $E(K, x)$ denoting encryption with the DES algorithm.

Because the key length of the DES algorithm is 56 bits and its block size is 64 bits, one would assume that an attacker has to apply an effort of the magnitude of around 2^{55} to discover key K and thus break the MAC, i.e., be in a position to compute valid MACs for new messages.

Security expectations

Unfortunately, the scheme is insecure because it allows attackers who do not know key k to compute valid MACs to construct messages. Let us assume that an attacker called Eve has recorded a message $(m, C_K(m))$ from Alice to Bob and that this message is 'secured' with a secret key K known only to Alice and Bob. Eve can use this message to construct a new message m' that is mapped to the same MAC:

The scheme is actually insecure!

❏ Let $y_1, y_2, ..., y_{n-1}$ be $n-1$ arbitrary 64-bit blocks.

❏ Eve computes the following for these values:

$$y_n := y_1 \oplus y_2 \oplus ... \oplus y_{n-1} \oplus \Delta(m)$$

❏ The new message m' is: $m' := (y_1, y_2, ..., y_n)$

❏ When Bob receives and checks the message $(m', C_K(m))$, he concludes that the contained MAC is correct and therefore accepts the message as an authentic message from Alice.

5.3.1   Schemes for the Computation of Message Authentication Codes

The two most frequently used schemes for computing message authentication codes break down into the following two categories:

- ❏ *'Cipher block chaining'-MACs (CBC-MACs)* use a block cipher in CBC mode and take the last ciphertext block as the message authentication code.

- ❏ *Message authentication codes based on MDCs* are used frequently because they are efficient, but cryptographically they should be viewed with some caution. This concerns some implicit though not yet proven assumptions about certain properties of the underlying MDC (also see Section 5.4).

Figure 5.4
General structure of a CBC-MAC

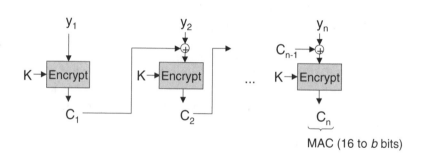

MAC (16 to *b* bits)

Little difference between computation of a CBC-MAC and CBC encryption

Figure 5.4 illustrates the general structure of a CBC-MAC. A CBC-MAC differs little from CBC encryption except that it requires the last block of the ciphertext even though it does not always use all of it. Because secret key K is already adopted into the computation of a MAC, it does not have to be encrypted again and can be used directly to prove the authenticity of a message.

MACs on their own cannot realise digital signatures

However, the the same key K must also be known to the checking entity. A MAC in principle therefore cannot be used to realise digital signatures that can be checked by independent third parties unless it deploys an additional asymmetric cryptographic algorithm.

5.3.2   Attacks on Message Authentication Codes

Regarding the security of message authentication codes based on block ciphers in CBC mode, note that attackers must obtain knowledge of pairs of form $(m_i, MAC(x_i))$ because without knowing key K they will not be able to compute a valid pair $(m, h_K(m))$. This condition means that the MACs used can be shorter than MDCs without causing any reduction in security.

There is no known attack technique that enables attackers to carry out birthday attacks without knowledge of key K. Valid combinations of form $(m, h(m))$ cannot be created without the correct key, unlike with cryptographic hash functions. On the other hand, if an attacker knows the key, he or she no longer needs to compute message authentication codes for arbitrary messages. This attack technique is therefore of little importance with respect to message authentication codes.

Birthday attacks do not function with MACs

Another approach that can be used to make CBC-MAC more secure is triple encryption (e.g. 3DES) with different keys on the last ciphertext block. It takes only two additional encryption operations to triple the effective key length, thus merely an additive constant increase of computational effort.

Multiple encryption increases security effectively

5.4 Hybrid Schemes

This section concludes the discussion in this chapter by considering hybrid schemes that try to implement either an MDC based on a MAC or a MAC based on an MDC.

A number of schemes were proposed that involved setting key K at a fixed value so that a scheme for computing modification detection codes could be obtained from one computing message authentication codes. However, the short block length of most common block ciphers (64 or 128 bits) means that such approaches do not offer adequate security against birthday attacks. Yet the parameterisation of block ciphers with key K is what ultimately makes CBC-MAC and related schemes less vulnerable to birthday attacks. Block ciphers are also normally much more computation-intensive than cryptographic hash functions and this inefficiency already gives them a considerable disadvantage over 'real' cryptographic hash functions.

MACs as MDCs

The reverse effort of using cryptographic hash functions to compute message authentication codes is mainly motivated by the fact that cryptographic hash functions require less computation. Products that only contain cryptographic hash functions and no real encryption algorithms are normally excluded from export restrictions, which is a further advantage.

MDCs as MACs

The basic idea behind the construction of MACs based on

Basic idea

MDCs is that a shared secret is 'mixed' with a message in such a way that an attacker is not able to construct an appropriate 'MDC-MAC' for a new message m' by eavesdropping on 'MDC-MACs' $h(m_i)$ for a range of messages m_i.

Security concerns   The assumption that an attacker must have knowledge of secret K in order to construct a valid 'MDC-MAC' into a new message m' raises some concerns regarding cryptography:

❏ For example, it has been shown that the construction $h(K \parallel m)$ is not secure, because an attacker can lengthen eavesdropped messages and then give these modified messages a valid 'MDC-MAC' without knowledge of key K [MOV97, Note 9.64].

❏ The construction $h(m \parallel K)$ is not secure for similar reasons [MOV97, Note 9.65].

❏ The construction $h(K \parallel m \parallel K)$, sometimes also called *prefix-suffix mode*, also does not offer adequate security [MOV97, Note 9.66].

HMAC   The following construction has established itself as the most popular scheme:

$$MDC - MAC(K, m) := h\big(K \parallel p_1 \parallel h(K \parallel p_2 \parallel m)\big)$$

It entails the use of two different 'padding patterns' p_1 and p_2 to fill up a key to the input block length of the cryptographic hash function. No known effective attacks have been made against this scheme, so it is currently regarded as secure [MOV97, Note 9.67]. The scheme was standardised in RFC 2104 ('Request for Comments') [KBC97] by the *Internet Engineering Task Force (IETF)* and is generally referred to as the *Hashed Message Authentication Code (HMAC)*

5.5 Summary

Delineating error   Cryptographic check values enable messages to be checked to en-
checksums   sure that they have not been intentionally modified after a check value has been created. In contrast to error checksums, such as Cyclic Redundancy Check (CRC), the effectiveness of cryptographic check values is not based on the non-applicable assumption that modifications to a message can only occur as the result of random events.

There are two principal categories of cryptographic check values: *Modification detection codes (MDC)*, which only receive the protected message as input and merely compute a 'digital fingerprint' of the message, and *message authentication codes (MAC)*, which are additionally parameterised with a secret key such that they can be directly used to verify the authenticity of the message.

Categories

Modification detection codes are computed with *cryptographic hash functions* and the two most commonly used algorithms are *MD5* und *SHA-1*. Because of the inadequate length of the output value of MD5 (128 bits) and a weakness in its internal compression function, MD5 should not be used for applications that require *collision resistance*. Collision resistance implies that it should not be possible to find two arbitrary messages that are mapped to the same modification detection code. This requirement together with an adaptation of the *birthday phenomenon* (only 23 people have to be in a room so that the probability of at least two people having the same birthday is 50%) accounts for the minimum length of the output value of cryptographic hash functions, which is currently estimated at 160 bits.

MDCs

CBC-MAC has established itself as a popular algorithm for computing message authentication codes. It uses the last ciphertext block with a block cipher as the message authentication code for encryption in 'cipher block chaining' mode.

MACs

Attempts at constructing modification detection codes based on message authentication codes have so far proven to be insecure and also too inefficient. With the reverse effort of constructing MACs based on MDCs, the *HMAC construction* is recognised as a viable scheme because it can be computed efficiently, is still considered secure and it is not subject to any export restrictions.

Hybrid schemes

5.6 Supplemental Reading

[KBC97]   KRAWCZYK, H.; BELLARE, M.; CANETTI, R.: *HMAC: Keyed-Hashing for Message Authentication*, February 1997. RFC 2104

[Mer89]   MERKLE, R.: One Way Hash Functions and DES. In: *Proceedings of Crypto '89*, Springer, 1989

[MOV97]   MENEZES, A.; VAN OORSCHOT, P.; VANSTONE, S.: *Handbook of Applied Cryptography*. CRC Press LLC, 1997

[Riv91]   RIVEST, R. L.: The MD4 Message Digest Algorithm.
           In: *Advances in Cryptology — Crypto '90 Proceedings*,
           Springer, 1991, pp. 303–311

[Riv92]   RIVEST, R. L.: *The MD5 Message Digest Algorithm*,
           April 1992. RFC 1321

[Rob96]   ROBSHAW, M.: *On Recent Results for MD2, MD4 and
           MD5*. November 1996. – RSA Laboratories' Bulletin,
           No. 4

[Yuv79]   YUVAL, G.: How to Swindle Rabin. In: *Cryptologia*
           July 1979

5.7 Questions

1. Delineate cryptographic check values from error checksums.
    Can cryptographic check values be used as error checksums?

2. Explain the difference between modification detection codes
    and message authentication codes.

3. Describe the generic structure of the two functions MD5 and
    SHA-1.

4. Explain the connection between the birthday phenomenon
    and cryptographic hash functions.

5. For SHA-1 give the value of W_{19}, processing the first message
    block.

6. Formulate the MD5 algorithm in programming language
    pseudo-code.

7. Why are message authentication codes not vulnerable to at-
    tacks based on the birthday phenomenon?

6 Random Number Generation

Generating random numbers for cryptographic algorithms is a subject that has appeared frequently in previous chapters (see, for example, Sections 4.3, 4.5 and 4.6). This chapter looks at how such 'random' calculations of numbers can be realised using principally deterministic algorithms.

6.1 Random Numbers and Pseudo-Random Numbers

Even though computers are excellent for all sorts of computation in a diverse range of application scenarios, in principle they are not really designed to generate truly 'random' results. Therefore, the seemingly trivial task of generating random numbers is more of an effort for a computer than one would think.

This is because algorithms and the hardware on which they are executed basically work in a deterministic way, and in reality random results cannot be generated through totally deterministic means.

In principle hardware and software work in a deterministic way

Consequently, the usual approach is to generate a large set of 'pseudo-random' information from a relatively small set of 'truly random' information. Before concrete methods are introduced to address this task, some concrete definitions are used to explain the exact difference between 'random' and 'pseudo-random'.

Definition 6.1 *A **random bit generator** (RBG) is a device or a scheme that outputs a sequence of statistically independent and uniformly distributed bits.*

Random bit generator

A random bit generator can be used to generate uniformly distributed random numbers in a given interval $[0, n]$. First a random bit sequence of length $\lfloor lg_2(n) \rfloor + 1$ is generated and converted into a positive integer. If the recovered number is greater than n, it is discarded and the process is repeated until a random number $\leq n$ is produced.

Pseudo-random bit
generator

Definition 6.2 *A* **pseudo-random bit generator** *(PRBG) is a deterministic algorithm that receives as input a truly random binary sequence of length k and produces as output a binary sequence of length m, which appears to be random and is distinctly longer than the input binary sequence. This input is referred to as* **seed** *and the established term for the output is* **pseudo-random bit sequence**.

The output of a pseudo-random bit generator is not really random. In actual fact, only a small number of 2^k outputs are generated out of a total of 2^m possible bit sequences because a generator always produces the same output for each fixed seed.

'True' randomness is
time consuming to
produce

The motivation for using pseudo-random bit generators is that it can be too time-consuming to produce truly random numbers of length m. Sometimes the only method available is manual like flipping a coin, and yet some systems (e.g. web servers that are supposed to support secure connections using SSL, see Chapter 12) require a large number of random numbers. Consequently, to an extent, a pseudo-random bit generator is used as an 'extension' of the 'expensive' truly random numbers used to initialise it.

Example of a PRBG

A simple example of a PRBG is the generator below, which generates a random bit sequence $y_1, y_2, \ldots$ according to the following linear recursion equation which is parameterised with three values a, b, q and initialised with a seed y_0:

$$y_i = a \cdot y_{i-1} + b \text{ MOD } q$$

6.2 Cryptographically Secure Random Numbers

The PRBG discussed above has a drawback that makes it unsuitable for cryptographic applications: even without knowledge of the values of a, b, q and y_0, attackers can use the observed values $\{y_i, y_{i+1}, \ldots, y_{i+j}\}$ to predict the other values $\{y_{i+j+1}, y_{i+j+2}, \ldots\}$ produced by a generator.

This section therefore defines the criteria for preventing this kind of undesirable prediction and for evaluating cryptographically secure random numbers.

The following requirements normally apply to pseudo-random bit generators that are used in cryptographic applications:

❏ Length k of the seed for a PRBG should be large enough so that a complete search of all possible seeds is not practical for an attacker using the resources available in our universe.

Requirements of cryptographic applications

❏ Statistically it should not be possible to distinguish the output of a PRBG from 'truly random' bit sequences (see Section 6.3).

❏ An attacker with limited resources should not be able to predict the output bits of a PRBG if he or she does not know the seed.

The last requirement mentioned, non-predictability of a pseudo-random bit sequence, is stated somewhat more formally in the following definition:

Definition 6.3 *A PRBG is acknowledged as having passed all statistical polynomial-time tests if no algorithm with polynomial-time exists that can correctly distinguish between an output sequence of the PRBG and a 'truly random' bit sequence of the same length with probability significantly greater than* 0.5.

'All polynomial tests' criterion

Polynomial-time in this case means that the runtime of the algorithm is bound by a polynomial of length m of the bit sequence.

Algorithms where the property mentioned can be proven obviously satisfy the security requirements introduced above. In practice however proof of this criterion turns out to be rather difficult.

The following criterion has proven to be much more manageable:

Definition 6.4 *A PRBG is acknowledged as having passed the next-bit-test if there is no polynomial-time algorithm that upon input of the first m bits of the bits generated by the PRBG can predict the $(m + 1)th$ bit as the output sequence with probability significantly greater than* 0.5.

Next-bit test

Although this definition clearly limits the set of algorithms that have to be considered compared to the first definition, it does show that both definitions describe the same PRBG (refer to [Sti95, Section 12.2] for the proof):

Theorem 6.1 *(Universality of the next-bit test) A PRBG passes the next-bit test when it passes all statistical polynomial-time tests.*

This background enables us to formulate the conditions under which a PRBG is considered secure for cryptographic applications:

CSPRBG

Definition 6.5 *A PRBG, which passes the next-bit test — possibly under some plausible but unproven mathematical assumption, such as the intractability of the factoring problem for large integers — is called a* **cryptographically secure pseudo-random bit generator***(CSPRBG).*

6.3   Statistical Tests for Random Numbers

It is recommended that several statistical tests be conducted on the output of a random bit generator, so that a certain confidence can be gained in the randomness of the generated bit sequences before the generator is used. A selection of different tests are listed below with a brief description:

Monobit test

❏ The *monobit test* verifies whether a bit sequence contains the same number of '1' bits as '0' bits.

Serial test

❏ The *serial test*, sometimes also called *two-bit test*, extends the idea of the monobit test to pairs of bits and checks whether a bit sequence contains an equal number of '00', '01', '10' and '11' pairs.

Poker test

❏ The *poker test* examines the number of sequences n_i of length q that contain the same value, with the parameter q of this test selected according to the length m of the analysed bit sequence such that $\lfloor m/q \rfloor \geq 5 \cdot 2^q$. Here too, random bit sequences should always produce equally distributed values.

Runs test

❏ The *runs test* checks whether the number of sequences containing either only '1' bits or only '0' bits ('runs') and being analysed for a range of differently specified sequence lengths fall within the numbers of 'runs' one would expect for 'truly random' bit sequences.

Autocorrelation test

❏ The *autocorrelation test* checks a bit sequence to verify whether correlations exist between the bit sequence and non-cyclically shifted (by a certain number of bits) versions of the bit sequence.

Maurer's universal test

❏ The *universal test from Maurer* checks whether bit sequences generated by a random bit generator can be compressed since compression of 'truly random' bit sequences is usually impossible.

The descriptions given here only highlight the basic idea of each test. The reader is referred to [MOV97, Sections 5.4.4 and 5.4.5] for a detailed and mathematically thorough description of the tests.

6.4 Generation of Random Numbers

Hardware- and software-based schemes, some outlined briefly below, are proposed for introducing 'true randomness' into the process of key creation.

Hardware-based random bit generators are normally based on the non-exact reproducibility of certain physical phenomena. Examples include:

Hardware-based random bit generators

❑ Observing the elapsed time between the emission of particles during radioactive decay [Gud85, Gud87].

❑ Measuring thermal noise [Ric92].

❑ Measuring the charge difference between two closely adjacent semiconductors [Agn88].

❑ Observing frequency fluctuations due to the instability of a freely running oscillator [AT&86].

❑ Measuring the amount by which a semiconductor can be charged during a fixed period of time.

❑ Observing the fluctuation in access times due to air turbulence in sealed disk drives.

❑ Recording sound using a microphone, or video input from a camera, e.g., at a busy intersection.

Manually throwing dice is also recommended as a secure method for generating what are called *master keys* in key hierarchies [MM78]. Due to the increase in the number of random numbers that need to be generated, this method is considered too time-consuming and therefore no longer up-to-date.

A hardware-based random bit generator should ideally also be implemented in a tamper-proof module to shield it from potential attackers.

Software-based random bit generators could be based on the following ideas:

Software-based random bit generators

❑ determining the current system time;

❏ measurement of elapsed time between keystrokes or mouse movements;

❏ reading the content of input and output buffers;

❏ use of user input;

❏ calculation of current values of an operating system, such as system load or statistics for network modules.

Combination of multiple random sources Ideally, multiple 'random sources' should be mixed together by concatenating their values and computing a cryptographic hash value to prevent attackers from being able to guess the random value. If the current system time is the only random source, then attackers can calculate the pseudo-random bit sequence that was computed on the basis of the seed if they have an approximate idea of when the pseudo-random bit generator was initialised.

Eliminating uneven distribution In case a random bit generator produces an uncorrelated but not uniformly distributed bit sequence, i.e., it produces '1' bits with the probability $p \neq 0.5$ and '0' bits with the probability $1 - p$, with p not necessarily a known but a fixed parameter, the following method can be used to eliminate the uneven distribution of '0' and '1' bits:

❏ the bit sequence is divided into pairs;

❏ all '00' and '11' pairs are discarded;

❏ the output is a '1' for each '10' pair and a '0' for each '01' pair.

On the assumption that the original bit sequence has no correlations, this scheme produces uncorrelated bit sequences in which '0' and '1' bits occur with the same frequency.

Eliminating uneven distribution using cryptographic hash functions An alternative practical method for eliminating uneven distribution or even correlations is the application of a cryptographic hash function to the output of the random bit generator. Even if the correctness of this method cannot be proven, it has been successfully used in practice and no known security flaws have yet been found.

6.5 Generating Secure Pseudo-Random Numbers

A number of different methods that use cryptographic hash functions or encryption algorithms have been proposed for generating cryptographically secure pseudo-random numbers. Although the security of these methods cannot be proven, they appear to offer sufficient security for most practical applications, and so far no security flaws have been discovered.

One representative of this class of methods is the generator *ANSI X9.17* that was standardised by the American Standardization Institute (ANSI) (see Figure 6.1).

```
// Algorithm: RandomX917
// Input:     a 64-bit long seed s,
//            an integer m,
//            a 3DES key k
// Output:    an m block long random bit sequence

int RandomX917(BitString64 s, int m, DES3_Key k)
{
  BitString64 q, x[];
  int         i;

  x = malloc(m*sizeof(BitString64));
  q = DES3(k, DateTime());
  for (i = 0; i < m; i++)
    { x[i] = E(k, Xor(q, s));
      s = E(k, Xor(x[i], q));
    } // of <for i>
  return(x);
}
```

Figure 6.1
The random generator ANSI X9.17

The ANSI X9.17 generator was certified for the pseudo-random generation of keys and initialisation vectors for use with DES and standardised as the 'U.S. Federal Information Processing Standard' (FIPS).

As mentioned above, it has not been proved that pseudo-random generators based on cryptographic hash functions or encryption algorithms are secure in the sense of Definition 6.5. Therefore, two schemes where this can be proven are presented below.

The first of these two schemes is the *RSA generator*, which, as its name suggests, is based on the RSA algorithm. The security of this generator stems from the difficulty of the factorisation problem.

RSA generator

❏ Output: a pseudo-random bit sequence $z_1, z_2, ..., z_k$ of length k

❏ Initialisation:

    ❏ generate two secret prime numbers p and q that are suitable for the RSA algorithm

    ❏ compute $n = p \cdot q$ and $\Phi = (p - 1) \cdot (q - 1)$

    ❏ choose a number e so that $1 < e < \Phi$ and $\operatorname{ggt}(e, \Phi) = 1$

    ❏ choose a random integer y_0 so that $y_0 \in [1, n]$

❏ For (i = 1; i $\leq$ k; i++)

    ❏ $y_i = (y_{i-1})^e \operatorname{MOD} n$

    ❏ $z_i =$ the lowest value bit of y_i

Improving efficiency   The efficiency of this generator can be slightly improved if j lower-valued bits of each number y_i are chosen, with $j = c \cdot lg(lg(n))$ guaranteed for a certain constant c that depends on the bit length m of the number n. However, so far it has not been possible to make a general calculation of the value of this constant for arbitrary m in such a way that the condition can be proven in Definition 6.5 (nevertheless for $j = 1$ the condition is met).

    The computational effort a generator requires to calculate one or j pseudo-random bits is considerable because an exponentiation with the number e modulo n has to be executed each time.

Blum-Blum-Shub   The *Blum-Blum-Shub generator*, named after its inventors, is
generator   a slight improvement in this respect. This generator replaces the exponentiation with e through squaring and its security is also based on the difficulty of the factorisation problem.

❏ Output: a pseudo-random bit sequence $z_1, z_2, ..., z_k$ of length k

❏ Initialisation:

    ❏ generate two secret prime numbers p and q so that p as well as q are each congruent to 3 modulo 4

    ❏ compute $n = p \cdot q$

    ❏ choose a number $s \in [1, n - 1]$ so that $\operatorname{ggt}(s, n) = 1$

    ❏ compute $y_0 = s^2 \operatorname{MOD} n$

❏ For (i = 1; i $\leq$ k; i++)

    ❏ $y_i = (y_{i-1})^2 \operatorname{MOD} n$

    ❏ $z_i =$ the lowest value bit of y_i

The efficiency of the Blum-Blum-Shub generator can be minimally increased in a similar way to the RSA generator if not one but j bits are chosen in each step, with similar conditions existing for the maximum number of bits allowed.

Improving efficiency

6.6 Summary

Generating random numbers is a requirement of many cryptographic protocols and therefore constitutes an important aspect of basic cryptographic knowledge. As only a small number of 'truly random bits' can usually be introduced into this process, *pseudo-random bit generators* that generate long pseudo-random bit sequences from a 'truly random' seed are normally used. It should not be possible to distinguish generated pseudo-random bit sequences from truly random bit sequences.

Note that not all pseudo-random bit generators are appropriate for cryptographic purposes as they also must satisfy the requirement of *non-predictability* if they are used for cryptographic applications. One way of proving that this prerequisite has been met is by showing that no algorithm with a polynomial time requirement can be constructed that can predict the $(m+1)$th output bit with probability greater than 0.5 when the first m output bits of the generator are known *(Next-Bit-Test)*.

Prerequisites for cryptographic purposes

Various hardware- and software-based methods are recommended for the generation of 'truly' random bits. However, before such methods are used, a series of statistical tests should be conducted on their output so that a certain confidence can be gained in the randomness of their output.

Random bit generators

In practice schemes based on cryptographic hash functions or encryption algorithms are a particularly efficient way of generating pseudo-random bit sequences. Although no theoretical proof of the security of these schemes exists, no serious security flaws have been discovered with them so far. The *RSA generator* and the *Blum-Blum-Shub generator* have been proven to be secure schemes for generating pseudo-random bit sequences but are relatively inefficient.

Pseudo-random bit generators

6.7 Supplemental Reading

[MOV97] MENEZES, A.; VAN OORSCHOT, P.; VANSTONE, S.: *Handbook of Applied Cryptography*. CRC Press LLC,

1997
Chapter 5 deals with generating cryptographically se-cure random numbers.

[Sti95]   STINSON, D. R.: *Cryptography: Theory and Practice (Discrete Mathematics and Its Applications).* CRC Press, 1995
Section 12.2 contains proof of the universality of the Next-Bit-Test (Theorem 6.1).

6.8   Questions

1.   Explain the differences between random numbers, pseudo-random numbers and cryptographically secure pseudo-random numbers.

2.   Why is initialisation of a pseudo-random bit generator with only current system time questionable from a security standpoint if it is to generate random numbers for use in (interactive) cryptographic protocols?

3.   Assume that you should be checking the output of a pseudo-random bit generator with the monobit test and the serial test. Write pseudocode for the two appropriate functions, making the assumption that the preceding generated random bit sequence is stored in a field $r[1, n]$ of data type 'bit'. How do you have to change your procedures if the field is of data type 'octet'?

4.   Write pseudocode for a procedure that removes inequality distributions of '0' and '1' bits from a pseudo-random bit sequence stored in the array $r[1, n]$. To which length does the generated pseudo-random bit sequence then shrink in the worst case?

5.   Does multiple recursive use of the procedure described for reducing uneven distributions of '0' and '1' bits make sense?

6.   Assume that you are basically processing each generated random bit sequence using the procedure to reduce uneven distribution. By how much does a generated random bit sequence shrink in an average case if the random bit generator supplies 'truly random' bits?

7 Cryptographic Protocols

There was a brief mention in Section 2.1 that cryptographic protocols represent the applications of cryptographic algorithms. We start this chapter by introducing the basic properties of cryptographic protocols and then follow with a close analysis of their two main fields of application for network security: authentication and key management.

7.1 Properties and Notation of Cryptographic Protocols

Definition 7.1 *A **cryptographic protocol** is defined as a series of steps and message exchanges between multiple entities in order to achieve a specific security objective.*

Similar to general communication protocols, cryptographic protocols must incorporate a number of properties to qualify as protocols:

General properties of protocols

❏ Each entity involved in a protocol run must know the protocol as well as all required steps and message formats in advance.

❏ Each entity involved in a protocol run must be in agreement with executing the protocol in accordance with the definition.

❏ The protocol must be unambiguous, i.e., all steps must be well defined and there should be no possibility of misunderstanding the protocol run.

❏ The protocol must be complete and, in particular, a specific action must be specified for each possible situation.

Cryptographic protocols must fulfill an additional requirement in order to fulfil their purpose:

Additional requirement

❏ It should not be possible to do or learn anything other than what is specified in the protocol.

Applications A multitude of different cryptographic protocols are available and implement a variety of applications:

❑ *Key exchange:* The Diffie–Hellman scheme for key exchange presented in Section 4.5 is an example of a pure key exchange protocol that allows no conclusion about the authenticity of the partners involved in the exchange.

❑ *Authentication:* This proves the identity of the sender of a message *(data origin authentication / data integrity)* or of a communication partner *(entity authentication)*. The latter security service also requires a guarantee of the currentness of message exchanges (see below).

❑ *Combined authentication and key exchange:* Many authentication protocols sometimes integrate optional key exchange into the authentication process.

❑ *Secret splitting:* If information is to be distributed over multiple entities in such a way that only all entities together can reconstruct and read the information, then a protocol of this class can be used to distribute the information. This is called *secret splitting*. One conceivable application for this would be dividing a code sequence to open a safe.

In a general case, information can even be divided up among n entities in such a way that a minimum of $m \leq n$ entities is required to read the information. Cryptographic protocols with this property are also referred to as *secret sharing*.

❑ *Time stamps:* It is often important to be able to prove that a particular message was created or sent at a specific time. This task can be performed with a time stamp using a recognised and trustworthy entity (e.g. a notary). The message is presented in plaintext to the trustworthy entity, who stamps the time and signs it.

❑ *Blind signatures:* In some circumstances, it is not desirable to notify a third party of the content of a message even though secure confirmation is needed of the exact time that the message was created. Protocols for creating *blind signatures* are available for this situation.

❑ *Key escrow for third parties:* Protocols in this category ensure that certain authorised entities (e.g. prosecution authorities), and only those entities, can learn the identity of the key used to secure a communication. Note that the demand for access to keys is controversial and that a great deal of doubt exists about the security of such protocols from a technical view.

❏ *Proof of knowledge without revealing information:* If an entity *A* wants to prove to another entity *B* that it has certain information without revealing the actual information, it can execute a protocol with *B* to provide *zero knowledge proof.* The protocol also guarantees that *A* can only execute the protocol correctly if it really has the information concerned.

❏ *Secure digital elections:* This enables electronic elections while ensuring that attempted fraud will not remain unnoticed.

❏ *Electronic payments:* This is a particularly interesting applications field for cryptographic protocols because of its importance as the basis for future *E-Commerce* applications although it is an area where the potential for abuse is especially high.

The list above shows that the versions and applications of cryptographic protocols are as diverse as the security needs that exist in the real world. Two of the categories mentioned, in particular, have special relevance for network security: *authentication* and *key distribution.* Both categories will be covered in detail in this chapter. Readers who are interested in specific protocols for the other categories listed are referred to the extensive specialist literature available [MOV97, Sch96].

Authentication and key distribution are particularly relevant for network security

7.2 Data Origin and Entity Authentication

Definition 7.2 Data origin authentication *is the security service that enables verification of the originator and the validity of a message. Verification is therefore possible at a later time to check whether the content of a message or, more generally, of data is still exactly the same as created by its originator. A synonymous term for this is* **data integrity**.

Data origin authentication and data integrity

Data origin authentication and data integrity are sometimes introduced as two different security services, with the first one guaranteeing the secure identification of the originator of a message and the second one the validity of the message itself. However, this artificial split makes little sense. Without the identity of the originator of messages, data integrity is meaningless in security terms, because all non-inherently faulty dates (thus all dates except 30 February, a PDU with an erroneous CRC checksum, etc.) are *a priori* 'integer'. What is of more interest is whether the date being checked derives from an attacker or from the expected originator. The relationship of the data integrity service to cryptographic protocols is twofold:

❏ Some cryptographic protocols implement the data integrity service. Normally, they consist of one protocol step and therefore are not particularly 'interesting' from a technical view. For example, Alice could prove the integrity of her messages by encrypting them with her private key. Every other entity that knows Alice's public key and has the assurance that it really is Alice's public key can check the integrity of her messages by decrypting the encrypted message using Alice's public key. Alternatively, Alice could also compute a modification detection value/code (see also Chapter 5) for her messages and then sign them with her private key.

❏ Data integrity of messages exchanged in cryptographic protocols is often an important property, so to an extent it represents a basic building block of cryptographic protocols.

Entity authentication **Definition 7.3 (Entity authentication)** *is the security service that enables communication partners to carry out a forgery-proof verification of the identity of their peer entities.*

Entity authentication is actually the most fundamental security service as all other security services are based on forgery-proof identification of the entities in a system and therefore build on this security service.

Implementation possibilities   In general, different means are available to implement entity authentication:

❏ *knowledge*, such as passwords, etc.;

❏ *ownership*, for example, physical keys or access cards;

❏ *immutable properties*, thus biometric properties such as fingerprints, etc.;

❏ *location area* of an entity. For instance, customers rarely check the authenticity of tellers in the branches of a bank;

❑ *delegation of authentication check*, where the verifying entity accepts that another entity it trusts has already established authentication.

Because it is difficult or insecure to use the means listed here to make direct verification in communication networks, cryptographic protocols are needed for this purpose. These will be examined in detail during the course of this chapter.

First we will examine the general difference between data origin and entity authentication, because the latter requires more than just an exchange of authentic messages.

Difference between data origin and entity authentication

The reason is that when entity B receives authentic messages from another entity A, it cannot be sure whether A really sent the messages at this specific moment or whether an attacker E is replaying old, recorded messages from A. The currentness of messages is therefore especially important in cryptographic protocols, specifically in applications where authentication verification only takes place once at the beginning of a communication — for example, through the transmission of a PIN at connection set-up.

There are two principal means for checking the currentness of a message:

Currentness check

❑ *Time stamps* which require clocks or index numbers that are synchronised with predetermined accuracy.

❑ *Random numbers* that are exchanged in challenge response dialogues.

Most of the authentication protocols covered in this chapter also simultaneously exchange session keys to secure the communication relationship that follows authentication. With some of these schemes there is no proof of authenticity until the session key is used.

In terms of a general approach, two main categories of protocols for entity authentication are distinguished:

Main categories of entity authentication

1. With *arbitrated authentication* an arbiter, or trusted third party (TTP), is involved in the authentication dialogue. This is particularly important when proof that authentication actually took place is required at a later date. Another reason for including a trusted third party is the possibility of using symmetric cryptology to implement services that permit authentication between entities that have not yet agreed a secret key with one another. Examples include the method by *Needham–Schroeder* [NS78] or the more secure

Otway-Rees scheme [OR87] and the *Kerberos* authentication and authorisation protocol for workstation computer clusters [Koh89, KNT94, Bry88].

2. *Direct authentication* of two partners is carried out without including an independent entity and therefore provides no proof to third parties that authentication actually took place. Nevertheless, each entity participating in the communication can guarantee the certainty that the respective peer entity is authentic. An example of this is the three modular authentication protocols of international Standard X.509 [ITU93].

Advantages of an authentication entity

The following advantages can result from the use of an arbitrated authentication entity:

1. If subsequent conflicts occur, only an independent entity can prove to third parties that authentication actually took place.

2. With the application of symmetric cryptographic schemes an authentication entity can be used to reduce the number of keys needed in a network. If n possible communication partners should authenticate each other in arbitrary pair combinations using a symmetric encryption scheme, this will require $(n^2 - n)/2$ secret keys because each possible pair has to share one secret key. When an authentication entity is introduced, agreement of n secret keys between the communication partners and the authentication entity is sufficient.

3. When a large number of authenticating entities exist, it may still be necessary to include independent entities in the process, even if the authentication scheme is based on asymmetric cryptography. The problem in such cases is the authenticity of the public key. An attacker M should therefore be prevented from tricking other entities into accepting 'his' key as the key of another entity A. Use of a trusted certification authority that can confirm the authenticity of keys through a signature with its own private key is recommended.

   The advantage is that the certification authority only has to confirm the authenticity of the corresponding public key and sign the certificate once for each entity and is, therefore, not actively involved in each authentication process. Consequently, authentication protocols of this type are often classified in the category of direct authentication.

Direct schemes based on symmetric cryptographic algorithms prove to be unsuitable when a large number of authenticating entities is involved. Because of the number of keys that have to be kept secret, secure distribution of the keys is no longer practical when authentication relationships between arbitrary pairs are to be established by entities. With methods based on symmetric cryptography with trusted entities a bottleneck tends to develop as the number of authenticating entities increases because of their active involvement in each authentication process. The trusted entity also can monitor all authentication activities, which in some circumstances can be undesirable.

Authentication based on symmetric cryptography is not suitable for very large populations

Even authentication schemes based on asymmetric encryption require additional certification of public keys by a trusted entity when a large number of authenticating entities is involved. This is the only practical way of implementing authentication and key management.

Requirements of key certification

The key representatives of both categories, the Needham–Schroeder protocol, the Otway–Rees protocol, Kerberos and international Standard X.509, are explained in the following sections. The notation used in the course of this discussion is summarised in Table 7.1.

7.3 Needham–Schroeder Protocol

The Needham–Schroeder protocol [NS78] enables two entities Alice (A) and Bob (B) to authenticate each other making use of a TTP entity and to negotiate a session key at the same time.

For this purpose the protocol uses a symmetric encryption algorithm as the fundamental cryptographic primitive. The trusted entity has a database with all users U that want to use the authentication service offered by the entity, as well as a secret key $K_{U,TTP}$ for each user.

The objective of a protocol is to enable two users, A and B, to obtain secure verification of each other's identity and, at the same time, to negotiate a session key for securing the communication that directly follows. The protocol consists of the following steps:

Protocol objectives

1. Alice chooses a random number r_A, creates a message that contains her name A, Bob's name B and the random number, and then sends this message to the trusted entity TTP:

$$A \rightarrow TTP: \quad (A, B, r_A)$$

Notation	Meaning
A	Name of A, analogous for B, TTP and CA
CA_A	Certification authority of A
r_A	Random number chosen by A
t_A	Time stamp generated by A
$(m_1, \ldots, m_n)$	Concatenation of messages m_1 to m_n
$A \rightarrow B : m$	A sends message m to B
$K_{A,B}$	Secret key only known to A and B
$+K_A$	Public key of A
$-K_A$	Private key of A
$\{m\}_K$	Message m encrypted with key K
$H(m)$	MDC over message m
$A[m]$	Shorthand notation of $\left(m, \{H(m)\}_{-K_A}\right)$
$Certificate_{-K_{CA}}(+K_A)$	Certificate for $+K_A$ issued by CA
$CA <\!< A >\!>$	Shorthand notation for $Certificate_{-K_{CA}}(+K_A)$

Table 7.1
Notation of cryptographic protocols

2. The trusted entity TTP generates a session key $K_{A,B}$ to secure the communication between A and B, encrypts this key together with the name of A using key $K_{B,TTP}$, which is agreed with B, and sends the following message encrypted with key $K_{A,TTP}$ to Alice:

$$TTP \rightarrow A: \ \left\{r_A, B, K_{A,B}, \{K_{A,B}, A\}_{K_{B,TTP}}\right\}_{K_{A,TTP}}$$

3. Alice decrypts this message, verifies that the random number r_A contained is the same one as in her first message and then sends the following message to Bob:

$$A \rightarrow B: \ \left\{K_{A,B}, A\right\}_{K_{B,TTP}}$$

4. Upon receipt and decryption of this message, Bob generates a random number r_B, encrypts it with $K_{A,B}$ and sends it to Alice:

$$B \rightarrow A: \ \left\{r_B\right\}_{K_{A,B}}$$

5. Alice decrypts the message with $K_{A,B}$, computes $r_B - 1$, encrypts the result with $K_{A,B}$ and sends it back to Bob:

$$A \rightarrow B: \ \left\{r_B - 1\right\}_{K_{A,B}}$$

6. Bob decrypts the received message and verifies that it contains $r_B - 1$. If it does, Bob assumes that he is really communicating with Alice.

The purpose of the last two messages is to allow Alice to prove to Bob that she really holds key $K_{A,B}$. Without knowledge of this key she would not be able to compute the response $\{r_B - 1\}_{K_{A,B}}$. As Bob knows that TTP also encrypted the session key $K_{A,B}$ in his message using Alice's key $K_{A,TTP}$, he concludes that Alice knows key $K_{A,TTP}$ and is therefore his authentic communication partner. However, this argumentation contains an error that can be exploited by attacker Eve (E) if she can discover a valid session key $K_{A,B}$ [DS81]. As the protocol offers no possibility of recognising an old session key from an earlier session, attacker Eve can impersonate as Alice to Bob if she has knowledge of an old session key $K_{A,B}$ and of the third message in the relevant protocol run:

Argumentation about the functioning of the protocol

Weakness of the protocol

1. Eve sends the recorded message:

$$E \to B: \ \{K_{A,B}, A\}_{K_{B,TTP}}$$

2. Upon receipt and decryption of this message, Bob generates a random number r_B, encrypts it with $K_{A,B}$ and sends it to Alice:

$$B \to A: \ \{r_B\}_{K_{A,B}}$$

3. This message is intercepted by Eve who decrypts it with $K_{A,B}$. Eve then computes $r_B - 1$, encrypts the result with $K_{A,B}$ and sends it to Bob:

$$E \to B: \ \{r_B - 1\}_{K_{A,B}}$$

4. Bob decrypts the received message and verifies that it contains $r_B - 1$. If it does, Bob assumes that he is really communicating with Alice, in which case Eve has successfully impersonated Alice.

Note that the protocol should actually be ensuring that only one entity with knowledge of key $K_{A,TTP}$ can impersonate Alice. However, due to the protocol flaw described, it is sufficient to have knowledge of a session key $K_{A,B}$ and the third message of the associated protocol in order to pass oneself off as Alice to user B. Since key $K_{A,B}$ could be used to encrypt a large quantity of data after a

Intended vs achieved protocol objectives

protocol run, it is potentially more open to cryptanalysis than key $K_{A,TTP}$. In summary, the Needham–Schroeder protocol provides less security than originally intended.

As a result, a number of cryptographers submitted proposals for improvements that would help the protocol meet its original objectives.

Otway–Rees protocol   In principle, the solution worked out by Needham and Schroeder themselves [NS87] is the same one as the *Otway–Rees protocol* [OR87] that appeared in the same journal:

1. Alice creates a message with a index number i_A, her name A, Bob's name B and the same information plus a random number r_A, encrypted with the key $K_{A,TTP}$ that she agreed with the trusted entity TTP, and sends this message to Bob:

$$A \rightarrow B: \ \left(i_A, A, B, \{r_A, i_A, A, B\}_{K_{A,TTP}}\right)$$

2. Bob generates a random number r_B, encrypts it and i_A, A together with B using key $K_{B,TTP}$ that he agreed with TTP and sends the following message to TTP:

$$B \rightarrow TTP: \ \left(i_A, A, B, \{r_A, i_A, A, B\}_{K_{A,TTP}},\right.$$
$$\left.\{r_B, i_A, A, B\}_{K_{B,TTP}}\right)$$

3. Upon receipt of this message TTP decrypts the two message parts contained, generates a new session key $K_{A,B}$ and two encrypted messages, one for Alice and one for Bob and sends both to Bob:

$$TTP \rightarrow B: \ \left(i_A, \{r_A, K_{A,B}\}_{K_{A,TTP}}, \{r_B, K_{A,B}\}_{K_{B,TTP}}\right)$$

4. Bob decrypts his part of the message using key $K_{B,TTP}$, verifies that the random number r_B contained matches the one he sent in his previous message and then sends Alice her part of the message:

$$B \rightarrow A: \ \left(i_A, \{r_A, K_{A,B}\}_{K_{A,TTP}}\right)$$

5. Alice decrypts this message using key $K_{A,TTP}$ and compares the contained random number r_A with the number generated in the first protocol step. Now when she uses session key $K_{A,B}$ for encrypted communication with Bob, she can be sure that she is communicating with Bob because only TTP would have been able to create message part $\{r_A, K_{A,B}\}_{K_{A,TTP}}$, an a potential attacker Eve is not able to modify message part $\{r_A, i_A, A, B\}_{K_{A,TTP}}$ that was created in the first protocol step.

Using the same argumentation, Bob can conclude that he is communicating with Alice when he receives interpretable messages from her that are encrypted with the session key $K_{A,B}$.

It goes without saying that Alice as well as Bob must be able to rely completely on the correct role and honesty of a trusted entity TTP.

7.4   Kerberos

The *Kerberos* authentication protocol was invented during the late 1980s at the *Massachusetts Institute of Technology (MIT)* in Boston as part of Project *Athena*. Kerberos is an authentication and access control service for workstation clusters. The main objectives in the design of Kerberos were:

Main objectives of Kerberos

❑ *Security:* Neither passive nor active attackers should be able to impersonate someone else when accessing a service or be able to eavesdrop on the information needed to do so.

❑ *Reliability:* Because each use of a service requires prior authentication, the Kerberos service itself must be particularly reliable and always available.

❑ *Transparency:* Beyond the requirement of entering a password at the beginning of a session, the authentication process should be largely transparent to the user.

❑ *Scalability:* Kerberos must have the ability to support a large number of users, workstations, services and servers.

The fundamental usage scenario of Kerberos is a user Alice who wants to use one or more services that are provided by different servers $S1$, $S2$, ... connected over an insecure network. Kerberos covers the following security aspects of the scenario:

Usage scenario and security aspects

❑ *Authentication:* Alice first authenticates herself to an authentication server (AS) that issues her with a temporary permit to request access to services. This permit is called a *ticket-granting ticket (TGT)* and is comparable to a passport with a limited duration of validity (lifetime).

❑ *Access control:* Alice uses the TGT in a second step to receive service-specific access authorisation, e.g., for access to server $S1$ that offers printing and file services. The TGS verifies that Alice is authorised to have access to the service requested and if so responds with a *service-granting ticket (SGT)* for server $S1$.

❏ *Key negotiation:* The authentication server generates a session key for the communication between Alice and TGS and the ticket-granting server generates a corresponding session key for the communication between Alice and the service-specific servers. As with the Needham–Schroeder and Otway–Rees protocols, this key is also used for authentication purposes.

Protocol run   Figure 7.1 presents an overview of a Kerberos protocol run.

Initiation   1. User Alice logs onto her workstation and requests access to a particular service. From now on Alice is represented by the workstation in the Kerberos protocol. The workstation sends the first message with Alice's name A, the name of an appropriate ticket-granting server TGS and a time stamp t_A to the authentication server:

$$A \rightarrow AS: \ (A, TGS, t_A)$$

Output of TGT   2. The authentication server AS verifies in its user database that it knows Alice and from Alice's password, which is also stored in the user database, generates a key K_A. It then extracts the network address of Alice's workstation $Addr_A$ from the protocol data unit received from Alice, creates a ticket-granting ticket $Ticket_{TGS}$ and a session key $K_{A,TGS}$ and sends the following message to Alice:

$$AS \rightarrow A: \ \{K_{A,TGS}, TGS, t_{AS}, LifetimeTicket_{TGS},$$
$$Ticket_{TGS}\}_{K_A}$$

$LifetimeTicket_{TGS}$ refers to the maximum time period the ticket is valid and $Ticket_{TGS}$ is defined as follows:

$$Ticket_{TGS} = \{K_{A,TGS}, A, Addr_A, TGS, t_{AS},$$
$$LifetimeTicket_{TGS}\}_{K_{AS,TGS}}$$

User authentication   3. Upon receipt of this message, the workstation asks Alice to *and request for a SGT*   enter her Kerberos password, which it uses to compute key K_A. This key enables the workstation to decrypt the message. If Alice does not enter the correct (authentic) password, the key K_A is not computed correctly and consequently the decrypted message does not produce interpretable plaintext. The remaining steps of the Kerberos protocol then fail.

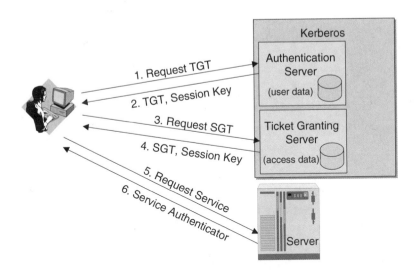

Figure 7.1
*Overview of the
Kerberos protocol*

Alice (thus the workstation) generates an *authenticator* and sends it together with her ticket-granting ticket and the name of desired server $S1$ to ticket-granting server TGS:

$$A \rightarrow TGS: \quad (S1, Ticket_{TGS}, Authenticator_{A,TGS})$$

The $Authenticator_{A,TGS}$ is defined as follows:

$$Authenticator_{A,TGS} = \{A, Addr_A, t'_A\}_{K_{A,TGS}}$$

4. The ticket-granting server TGS decrypts $Ticket_{TGS}$, extracts key $K_{A,TGS}$ from the resulting plaintext and uses it to decrypt $Authenticator_{A,TGS}$. If the name and address of the authenticator and the tickets match and the time stamp is still sufficiently valid, it then verifies that Alice is allowed access to server $S1$, produces a time stamp t_{TGS}, a session key $K_{A,S1}$, a ticket $Ticket_{S1}$ for access to server $S1$ and sends the following message to Alice:

Authorisation check and output of SGT

$$TGS \rightarrow A: \quad \{K_{A,S1}, S1, t_{TGS}, Ticket_{S1}\}_{K_{A,TGS}}$$

in which $Ticket_{S1}$ is defined as follows:

$$Ticket_{S1} = \{K_{A,S1}, A, Addr_A, S1, t_{AS}, LifetimeTicket_{S1}\}_{K_{TGS,S1}}$$

5. Alice decrypts this message and now holds a session key for secure communication with server $S1$. She generates a new authenticator and sends it together with her ticket to $S1$:

Authentication to server

$$A \rightarrow S1: \quad (Ticket_{S1}, Authenticator_{A,S1})$$

with $Authenticator_{A,S1}$ in turn defined as:

$$Authenticator_{A,S1} = \{A, Addr_A, t'_A\}_{K_{A,S1}}$$

Authentication check by server

6. Server $S1$ decrypts the received ticket using key $K_{TGS,S1}$ that was agreed between it and the ticket-granting server and thus obtains session key $K_{A,S1}$. It uses this key to verify the authenticator and then responds to Alice:

$$S1 \rightarrow A: \ \{t'_A + 1\}_{K_{A,S1}}$$

Verifying server authenticity

7. Alice decrypts this message and checks the contained time stamp incremented by one. If this verification is successful, Alice assumes she is communicating with server $S1$, because except for TGS, only $S1$ knows key $K_{TGS,S1}$ and can use it to decrypt $Ticket_{S1}$ that contains session key $K_{A,S1}$. Therefore, only $S1$ is able to decrypt $Authenticator_{A,S1}$ and to respond with the incremented time stamp $t'_A + 1$ encrypted with key $K_{A,S1}$.

Inter-realm authentication

This fundamental authentication dialogue can be extended to a protocol for *inter-realm authentication*. Let us look at an organisation that operates workstation computer clusters at two different locations. We will assume that user A at the first location wants to access the service of a server at the second location. If autonomous Kerberos servers and user databases are being operated at both locations, this means that actually two distinct domains exist, in Kerberos terminology also called *realms*. Kerberos enables inter-realm authentication to avoid a user A having to be registered in both domains. Figure 7.2 shows the entities involved and an overview of the protocol run.

Secret key required

Inter-realm authentication requires that the two ticket-granting servers of both domains have together agreed upon a secret key $K_{TGS1,TGS2}$. The basic idea is that the local ticket-granting server views the remote ticket-granting server as a 'normal' server and therefore can issue a ticket for it. After Alice obtains a ticket-granting ticket (TGT_{rem}) for the remote domain, she sends a request to the remote ticket-granting server to issue her with a service-granting ticket for requested server $S2$. With inter-realm Kerberos authentication it is crucial that the remote domain trusts the authentication server of the local domain as it does not carry out its own authentication check of 'visiting' users.

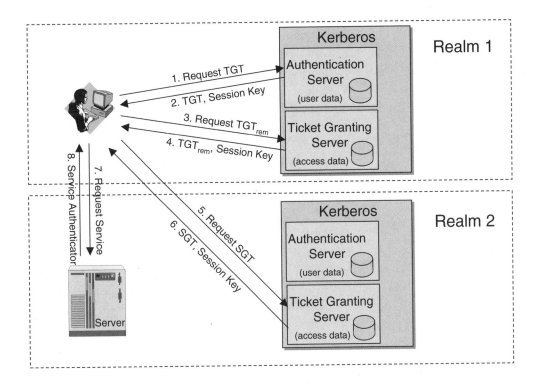

Figure 7.2
*The Inter-Realm
Kerberos Protocol*

The discussion so far has described Version 4 of the Kerberos protocol. A number of shortcomings were found in this version that can be classified into two categories:

❑ *Technical shortcomings*: Because Kerberos Version 4 was developed as part of Project Athena and with regard to the special requirements of this project, it is not orientated towards general usability. Consequently, the following shortcomings are evident:

    ❑ Dependence on a concrete encryption algorithm: Version 4 of the Kerberos protocol stipulates use of the DES algorithm.

    ❑ Dependence on a concrete communication protocol: The format of the address field $Addr_A$ is defined so that it can only accept IP addresses.

    ❑ Byte sequence in messages: The protocol uses a proprietary method for specifying the byte sequence.

    ❑ Period of ticket validity: Because the period of ticket validity is specified in an octet in units of 5 minutes, the maximum validity of a ticket is little more than 21

*Technical
shortcomings*

DES

Only IP-based

Byte sequence

*Period of ticket
validity*

hours. This time span is not sufficient for a variety of applications, such as extensive simulations, which need a valid ticket for the duration of their computation.

Lack of delegation

❏ Delegation of authentication/authorisation: Kerberos Version 4 does not support delegating authorisation, a facility that can be very useful in some circumstances. For instance, if delegation is allowed, a printing service can be given the right to access a certain file on behalf of user that wants to print the file.

Scalability

❏ Inter-realm authentication: The method of inter-realm authentication does not scale well for large numbers N of connected domains because basically $(N^2 - N)/2$ secret keys have to be generated and distributed if users from all domains should have access to servers of any domains.

Security flaws

❏ *Security flaws*: In addition to the technical shortcomings described, Kerberos protocol Version 4 also has some cryptographic flaws:

Double encryption

❏ Double encryption: The double encryption of the tickets in the second and fourth protocol step does not increase security and therefore is a waste of computing resources.

PCBC-Mode

❏ Encryption in PCBC-Mode: The non-standardised operating mode *Propagating Cipher Clock Chaining (PCBC)* was specified for encryption with DES. The intention was to implement confidentiality and data integrity into one step without the need to compute a modification detection code. Unfortunately, attack techniques against PCBC mode enabling the substitution of ciphertext blocks were discovered and therefore it does not provide adequate data integrity protection [Koh89].

Vulnerability to replay attacks

❏ Session key: Each ticket contains a session key that the user utilises to encrypt an authenticator. However, the same key is also used to secure the communication between workstation and server that immediately follows. The fact that a user can use a ticket for multiple independent service uses means that successive communication relationships are vulnerable to replay attacks. A better approach would be to have a strict separation between keys that are used for authentication and session keys that protect exchanged data units and always to negotiate new keys for the second use.

❏ Attacks on passwords: Because the actual authentication key K_A of a user A is generated directly from the password of the user, Kerberos is vulnerable to 'password guessing attacks' where attackers systematically try out different passwords. Another problem with Kerberos Version 4 is that the authentication server responds to the first unprotected message of a user with a message encrypted with key K_A. The general structure of this message is known and furthermore the user has contributed the content of it. Consequently, an attacker can easily find an encrypted message where he knows some of the parts and can use it for a systematic search for the key by trying out frequently selected passwords.

Vulnerability to password attacks

In reaction to the shortcomings that were discovered, a new version of the Kerberos protocol was specified, eliminating the criticisms described above [KNT94]. For space reasons this version will not be dealt with here and for a detailed description interested readers are referred to the literature mentioned.

Kerberos Version 5

7.5 International Standard X.509

In 1988 Recommendation X.509 [ITU93] was standardised by the *International Telecommunications Union* (ITU) as part of the X.500 recommendations for the provision of directory services. Use of the X.500-Directory Service for the realisation of authentication services is described in this recommendation and relevant procedures and data formats are specified. In terms of content, the recommendation makes a distinction between two independent parts:

1. As an important prerequisite for the practical application of asymmetric cryptographic algorithms for the realisation of security services, the recommendation describes methods for a worldwide and secure distribution of public keys. This includes the definition of *certificates* for public keys to prove their authenticity and methods for hierarchical certification.

Key certification

2. The recommendation also contains various schemes for authenticating entities. These range from 'simple' authentication based on passwords to 'strong' *authentication protocols*.

Authentication protocols

The concepts of the certificates based on X.509 are introduced briefly below. The section that follows describes the authentication protocols of the recommendation.

7.5.1   X.509 Key Certificates

Certificates are issued by a *certification authority* – in its function comparable to an identification card authority – and essentially contain the unique name of the 'holder', his or her public key, the period of validity and a digital signature. Figure 7.3 shows the structure of an X.509 certificate.

Figure 7.3
*Structure of X.509
certificates*

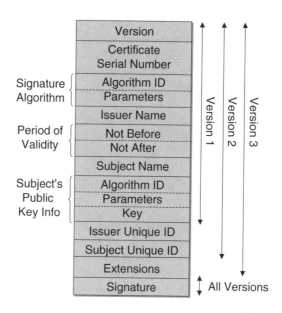

For clarity, the following notation will be used in the following explanations:

$$A[I] := (I, \{H(I)\}_{-K_A})$$

In the notation $-K_A$ denotes the private key of entity A, I arbitrary information for signing, H a cryptographic hash function and $\{x\}_{-K_A}$ the value of x encrypted with A's private key. In X.509 notation, a certificate issued by certification authority CA for the public key of entity A is structured as follows:

$$CA \ll A \gg := CA[V, SN, AI, CA, T_A, A, +K_A]$$

*Components of X.509
certificates*   It contains the version number V of the X.509 standard, one unique serial number SN per certification authority, the name and the parameters of the signature scheme AI used, the name of the certification authority CA, the period of validity of certificate T_A and the name of the entity being certified A along with its public

key $+K_A$ (also see Figure 7.3). CA signs this information with its private key $-K_{CA}$.

If the entities that want to authenticate each other know the public key $+K_{CA}$ of the certification authority, they can use the certificate for verification of each other's public key.

If the number of users requiring authentication is extensive or users are separated by large geographical distances, it is not practical for one central certification authority to sign all public keys. Consequently, X.509 provides for the formation of a *certification hierarchy*. In such a hierarchy additional certificates are used to link together multiple certification authorities. Figure 7.4 shows a hypothetical X.509 certification hierarchy (this illustration only shows certification authorities and no users).

Certification hierarchy

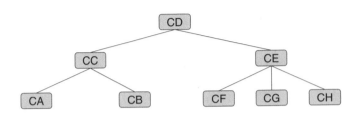

*Figure 7.4
Example of an X.509
certification hierarchy*

In this example an entity A that trusts certification authority CA can check the public key of entity H, whose key is certified by certification authority CH through a *chain of certificates*. In X.509 notation a certificate chain is structured as follows:

$$X_1 \ll X_2 \gg X_2 \ll B \gg$$

In this chain, certification authority X_2 is certifying entity B. X_2 in turn is certified by certification authority X_1. So that the chain can also be formed in the opposite direction, X_2 also has to certify X_1. In X.509 the length of certification chains is not limited. If entity A in the example shown in Figure 7.4 wants to verify the public key of entity H, she will need the following certification chain:

$$CA \ll CC \gg CC \ll CD \gg CD \ll CE \gg CE \ll CH \gg CH \ll H \gg$$

Because the certificates cannot be forged if the certification authorities are operated securely, they can be stored in a public directory. X.509 provides for the X.500 Directory Service that contains the appropriate ASN.1 definitions of the required data formats.

*Storage in public
directories*

As a rule, a new certificate is issued in time before the currently valid certificate expires. In some circumstances it may be necessary to invalidate a certificate even before it expires:

*Reasons for
premature
invalidation*

❏ There is a suspicion that the private key of an entity has become known.

❏ The issuing certification authority is no longer certifying a particular entity.

❏ There is a suspicion that the private key of a certification authority has become known. If this is the case, all certificates signed with this key must be invalidated.

Certificate revocation list

Therefore, in addition to certificates, the X.500 directory also stores what are called *certificate revocation lists* that contain all invalidated certificates. The verification of a certificate also has to check whether it has already been invalidated. The directory service is normally consulted for this information.

7.5.2 Direct Authentication Protocols Based on X.509

In addition to password-based and so-called 'simple' authentication, the X.509 recommendation defines three cryptographic authentication protocols that are referred to as 'strong' and build on one another. These are defined as follows:

One-way authentication

1. If entity A is merely being authenticated to entity B, then A sends the following message to B *(one-way authentication)*:

$$A \rightarrow B : (A[t_A, r_A, B, sgnData, \{K_{A,B}\}_{+K_B}], CA \ll A \gg)$$

This entity receives a message signed by A and the certificate $CA \ll A \gg$. The signed message contains a time stamp t_A, a unique and random value r_A within the period of validity of the time stamp, the name of B, optional user data $sgnData$ and a likewise optional session key $K_{A,B}$, which is encrypted with the public key $+K_B$ of B.

2. Upon receipt of this message, B checks the respective certificate with $+K_{CA}$ and extracts the public key $+K_A$ of A. To check the authenticity of the key, B may have to request other certificates from the public directory and verify a certification chain from its own certification authority all the way to the public key of A. Then B uses key $+K_A$ to verify the message computed by A through the received message. Based on the values t_A and r_A and B, B either accepts the authentication or declines it. The scheme requires that all participants have synchronised clocks. The value r_A of B is stored until the validity of time stamp t_A has expired so that the tolerated clock difference selected does not have to be too

small. Within this time span B accepts no authentication messages with the same r_A from A.

3. If mutual authentication is desired, then B sends a similar message to A *(two-way authentication)*:

$$B \rightarrow A: (B[t_B, r_B, A, r_A, sgnData, \{K_{B,A}\}_{+K_A}], CA \ll B \gg)$$

Two-way authentication

The time stamp contained in the message is not necessarily required as A can check whether the value r_A matches both the value sent by her earlier and the value generated for the current dialogue.

4. If no synchronised clocks are available, *three-way authentication* can take place. For this kind of authentication A additionally sends the following message to B:

$$A \rightarrow B: \{A[r_B]\}$$

Three-way authentication

In this instance A is authenticating herself to B by being able to sign the random number generated by r_B in the current dialogue.

Recommendation X.509 designates the use of asymmetric encryption algorithms for the digital signatures of messages. However, the authentication protocols described above can just as easily be executed with a signing scheme based on symmetric cryptography. In this case A and B must agree a secret authentication key $AK_{A,B}$ beforehand, in which case conveying and verifying certificates becomes superfluous.

The protocols can basically also be used with symmetric cryptography

7.6 Security of Negotiated Session Keys

The cryptographic protocols discussed so far in this chapter carry out entity authentication, simultaneously negotiating a session key for the communication relationship that immediately follows. More precisely, 'negotiation' of the key always involves one entity specifying a key and distributing it to the other entity or entities.

The Diffie–Hellman scheme for key negotiation was presented in Section 4.5 with an explanation that it does not offer its own authentication mechanism for message exchanges. This means that, without additional cryptographic measures, neither of the two entities can be certain about which entity it has exchanged a session key. This circumstance suggests linking Diffie–Hellman key exchange to supplementary data origin authentication. For example,

Lack of authentication with Diffie–Hellman

after or during key negotiation, each entity computes a signature over all the message elements it transmits and then sends it to the peer entity.

However, this requires access to an authentication key that both parties already have or to certified public keys. The question thus arises about the added benefit of using the Diffie–Hellman scheme if its security has to rely on other keys. After all, these authentication keys can just as easily be deployed directly to distribute a session key.

Advantages of key negotiation with Diffie–Hellman

Depending on the security requirements demanded of the key distribution, use of the Diffie–Hellman scheme is advantageous or necessary because it guarantees a security feature that cannot be provided when session keys are distributed directly. This is called *Perfect Forward Secrecy* .

Perfect Forward Secrecy

Definition 7.4 *A cryptographic protocol for key negotiation guarantees the property of* **Perfect Forward Secrecy** *when a possible compromising of a key $K1$ in the future will have no effect on the security of exchanged or stored data protected with a session key $K2$ that was negotiated with the help of key $K1$ before $K1$ was compromised.*

The cryptographic protocols presented in this chapter cannot guarantee Perfect Forward Secrecy. For example, with Kerberos authentication and key distribution, if the authentication key K_A becomes known at a later time, an attacker can subsequently decrypt all recorded messages by first computing session key $K_{A,TGS}$ from the response message of the authentication server to A and then use this key to ascertain session key $K_{A,S1}$ from the response message of the ticket-granting server to A.

Perfect Forward Secrecy can be guaranteed with the Diffie–Hellman scheme

If, however, two entities A and B negotiate a session key $SK_{A,B}$ using the Diffie–Hellman scheme and authenticate their messages with a secret authentication key $AK_{A,B}$, an attacker cannot subsequently use the authentication key to ascertain the session key because this would still at least require the computation of a discrete logarithm. The attacker also cannot actively attack the Diffie–Hellman scheme afterwards using a man-in-the-middle approach as the dialogue has already taken place and therefore can no longer be manipulated. Consequently, Diffie–Hellman is used for key negotiation in several network security protocols. These protocols are covered extensively in Part 2.

7.7   Formal Validation of Cryptographic Protocols

In the literature there are a number of known examples in which a cryptographic protocol contains flaws that enable attackers to influence the protocol without requiring the appropriate key, or where the cryptographic algorithms used by the protocol have to be broken [Mea95]. One example is the *Needham–Schroeder protocol* [NS78] that allows attackers to specify a potentially compromised key as the new session key (also see Section 7.3) [DS81]. Other examples are the authentication protocol of the draft version of international Standard *X.509* [ITU87], which contains a similar flaw [BAN90], and the *licensing system for software by Purdy, Simmons and Studier* [PSS82], which an attacker was able to circumvent by combining recorded messages [Sim85].

These examples motivate the need for a formal validation of cryptographic protocols because informal methods are not adequately able to analyse security flaws in protocols. An overview of popular approaches in this field is therefore given followed by a detailed description of a particularly successful approach, *GNY-Logic*.

7.7.1   Classification of Formal Validation Methods

A number of approaches were developed for the formal validation of cryptographic protocols. These can be divided into the following four categories [Mea92]:

1. *General approaches to prove certain protocol properties:* Approaches in this category consider cryptographic protocols as conventional programmes and analyse them using the usual methods of software verification, such as automaton-based approaches [Sid86, Var89] or predicate logic [Kem89]. Alternatively, they use special specification languages to describe and analyse protocols [Var90]. In its requirements, proof of the security of a cryptographic protocol differs considerably from proof of the correctness of a protocol, because malicious manipulation normally does not need to be considered in the latter proof. The approaches in this category are therefore not particularly suitable for analysing attacks on cryptographic protocols.

General approaches

Expert system based approaches

2. *Expert system based approaches:* Approaches of this type use the knowledge of human experts formalised with deductive rules, thereby enabling an automated and possibly interactive analysis of cryptographic protocols [LR92, MCF87]. Although this procedure lends itself well to the analysis of protocols with known attack patterns, it is less appropriate for tracking down loopholes based on unknown attack patterns [Sch96, p. 66].

Algebraic approaches

3. *Algebraic approaches:* Approaches in this category model cryptographic protocols as algebraic systems. In addition to the protocol steps, the model records the knowledge and the beliefs obtained by peer entities during an authentication dialogue and analyses the resulting model using algebraic operations. Examples of approaches in this category are provided by [Mer83, Tou91, Tou92a, Tou92b, WL93].

Logic-based approaches

4. *Special logic-based approaches:* This category was established as a result of the approach *BAN-Logic*, named after its inventors Burrows, Abadi and Needham [BAN90]. Since its introduction, it has maintained its reputation as the most successful strategy for a formal analysis of cryptographic protocols. Approaches of this type define a series of predicates, together with mapping instructions for converting message exchanges into formulas, thereby enabling an analysis of the *knowledge* and the *beliefs* that peer entities obtain during an authentication dialogue as part of the protocol run. Numerous improvements have been proposed for BAN-Logic and for extended approaches based on the same idea [GS91, GNY90, KW94, MB93, Oor93, Sne91]. Other logic-based approaches include [BKY93, Bie90, Mos89, Ran88, Syv90, Syv91, Syv93a, Syv93b, SO94].

GNY-Logic [GNY90] is considered to be one of the most successful of these approaches, and because of its general acceptance will be presented in detail.

7.7.2 GNY-Logic

GNY-Logic was named after its inventors *Gong, Needham* and *Yahalom* and published for the first time in 1990 [GNY90]. Analyses using this logic focus on systematically deducing the *knowledge* and *beliefs* obtained by the peer entities of a cryptographic protocol during a dialogue from previously specified conditions and message exchanges.

In this context, *knowledge* is knowing certain data values that are exchanged or negotiated during a dialogue (e.g. keys), whereas *beliefs* assess the exchanged data. Examples of beliefs include assumptions that a message was sent by a specific communication partner (and therefore not by an attacker), that a reported value is appropriate for use as a session key, or that a certain message is not a repetition of a message sent earlier.

The cryptographic protocol is assumed to be a distributed algorithm implemented by communicating state automatons. Message exchanges between automatons can be arbitrarily delayed, replayed, deleted or modified by attackers or even invented and inserted by them during transmission. It is also assumed that attackers cannot break the cryptographic algorithms used to secure message exchanges. Each partner of a cryptographic protocol manages two disjunct quantities:

1. *Knowledge*, the set of data values he or she holds or obtains during a dialogue.

2. *Belief*, the set of his or her beliefs about these data values.

Before a protocol run, these sets receive initial elements that are obtained from the preconditions of the authentication protocol. During the protocol new elements can be obtained for these sets from messages received. The following means of expression can be used for the notation of these elements:

❑ *Formulas* are used as names for variables that can contain arbitrary bit patterns. In the notation of GNY-Logic, X and Y refer to arbitrary data values, whereas K and S are used for keys and shared secrets. With X and Y, the union of these values (X, Y) is also a formula, with the union in the logic treated as a quantity with the properties associativity and commutativity.

In addition, $\{X\}_K$ denotes the value X symmetrically encrypted with key K and $\{X\}_K^{-1}$ denotes the plaintext associated with X and key K after encryption with a symmetric scheme. It therefore holds that $\{\{X\}_K\}_K^{-1} = X$.

$\{X\}_{+K}$ and $\{X\}_{-K}$ are used for the notation of asymmetric schemes and it holds that $\{\{X\}_{+K}\}_{-K} = X$.

$H(X)$ denotes the value that results from the application of a cryptographic hash function to the value X. Lastly, $F(X)$

denotes a function that can be reversed with a realisable effort, such as the XOR function.

Statements   ❑ *Statements* describe the knowledge and the beliefs of the peer entities of a protocol about certain formulas. Let P and Q be possible peer entities of a protocol and X, S and $+K$ be formulas. The following statements can then be formed:

Receipt of a message   ❑ $P \lhd X$: Entity P receives message X. Message X can be extracted from the received message either directly or by using a computation algorithm known to the communication partners.

Possessing a formula   ❑ $P \ni X$: Entity P possesses formula X. An entity possesses all initial elements of the quantity knowledge as well as all formulas that it obtains during a protocol run. In addition, it possesses all formulas that it can compute from existing formulas.

Transmitting a formula   ❑ $P \mid\sim X$: Entity P has transmitted X. Formula X can be directly transmitted in a message or computed from a message transmitted by P.

Currentness of a formula   ❑ $P \mid\equiv \#(X)$: Entity P has the belief that formula X is *new*, meaning that this formula was never used in earlier protocol dialogues. This property can be guaranteed by the fact that formula X was generated with a random generator.

Recognisability of a formula   ❑ $P \mid\equiv \phi(X)$: Entity P has the belief that it recognises formula X. It is possible that P has certain knowledge about the structure of formula X before it receives the formula. It is also possible that P recognises a specific value in the formula or that the formula contains a certain redundancy. This property is needed if an entity should be prevented from accepting a randomly generated value as a formula.

Confidentiality of a formula   ❑ $P \mid\equiv P \xleftrightarrow{S} Q$: Entity P believes that value S is an appropriate secret for securing message exchanges between P and Q. This secret can be a symmetric encryption key or a value for the signature using cryptographic hash functions. The notation is symmetric, which means $P \xleftrightarrow{S} Q$ and $Q \xleftrightarrow{S} P$ can be used synonymously.

Authenticity of public keys   ❑ $P \mid\equiv \xrightarrow{+K} Q$: Entity P has the belief that value $+K$ is the public key of Q and that the corresponding private key $-K$ is only known to Q.

C_1 and C_2 can also be used to form other statements:

❑ C_1, C_2: In GNY-Logic a union of statements is handled as a union of quantities with the properties associativity and commutativity.

Union of statements

❑ $P \mid\equiv C_1$: Entity P has the belief that statement C_1 is true.

Beliefs about statements

❑ $P \mid\Longrightarrow C1$: Entity P is competent to evaluate statement $C1$.

Competence

Furthermore, an X formula can have a '$\star$' prefix attached to it. In this case, entity P, which possesses the formula, knows that it has not produced and transmitted formula X in the current protocol dialogue itself. The statement $P \lhd \star X$ is an example of the use of this prefix. It is required in the analysis of cryptographic protocols to ensure that certain formulas can not be *reflected* by an attacker, meaning that they are sent back to the actual sender of a message.

In GNY-Logic the symbol '$\star$' is also used in another connection. The statement $P \mid\Longrightarrow P \mid\equiv \star$ means that P has the jurisdiction to judge all its beliefs.

In conclusion, a condition C can be attached to a formula X to express that an entity should only transmit formula X to a peer entity if the statement C is true. This fact is expressed in GNY-Logic by attaching symbol $\leadsto$ followed by statement C to a formula X: $(X \leadsto C)$.

In analyses with GNY-Logic, cryptographic protocols are first conveyed through syntactical analysis into statements of the form described above:

Transforming a protocol into statements

❑ Two lines $P \mid\sim X$ followed by $Q \lhd X$ are created for each line of the form $P \rightarrow Q : X$. The assumption is that the transmission service is reliable and that no errors that cannot be attributed to intentional manipulation will occur.

❑ After all messages of a protocol are processed, all lines of the form $P \mid\sim X$ and $P \lhd Z$ are analysed in the sequence of their occurrence for each partner P of the protocol. Each partial formula Y of a formula Z is checked in a statement $P \lhd Z$ to verify whether it occurred previously in a formula X of a statement $P \mid\sim X$. If this is not the case, then it means that P did not transmit the message itself and a '$\star$' prefix is attached to the beginning of partial formula Y. Once this step is completed, it is possible at a syntactical level to recognise whether a specific formula Y can possibly be reflected by an attacker. After this step all lines of form $P \mid \sim X$ are deleted.

❑ In a final step formulas of the form $\{\{X\}\}_K^{-1}$ are reduced to X. A corresponding reduction is made for asymmetrically encrypted values.

The actual analysis of the protocol takes place after these preparatory steps are completed. An attempt is made to use the deductive rules of GNY-Logic to deduce new formulas from the given formulas and those obtained during the dialogue until the objectives of the cryptographic protocol are derived.

Protocol objectives   Common objectives of cryptographic protocols are that all peer entities (e.g. P and Q) possess a specific key K after a dialogue is completed and believe in the secretness of this key: $P \ni K, Q \ni K,$ $P \mid\equiv P \xleftrightarrow{K} Q$ and $Q \mid\equiv P \xleftrightarrow{K} Q$. Another objective is that all peer entities believe that the other peer entities also believe in the secretness of the key: $P \mid\equiv Q \mid\equiv P \xleftrightarrow{K} Q$ and $Q \mid\equiv P \mid\equiv P \xleftrightarrow{K} Q$.

Rules of deduction of   GNY-Logic provides a set of rules of deduction to deduce these
GNY-Logic   objectives from the initial conditions and the formulas added during a protocol run. Some of these rules are listed below with an identifier chosen by the inventors of GNY logic indicated in parentheses:

(T1)   ❑ Receipt of formula X, in particular, can be inferred from the receipt of a formula $\star X$ that was not signed by P itself (Rule T1, 'T' denotes the 'being-told' rules on the receipt of formulas):

$$\frac{P \triangleleft \star X}{P \triangleleft X}$$

(T2)   ❑ Receipt of each of its partial formulas can be inferred from the receipt of a combined formula (T2):

$$\frac{P \triangleleft (X, Y)}{P \triangleleft X}$$

(T3)   ❑ If P receives a formula $\{X\}_K$ that is encrypted with key K and if P is in possession of key K, then this can be used to infer the receipt of formula X (T3):

$$\frac{P \triangleleft \{X\}_K, \ P \ni K}{P \triangleleft X}$$

❑ If P receives a formula X, then it can be concluded that P is   (P1)
in possession of formula X (P1, 'P' generally identifies 'possession' formulas):

$$\frac{P \lhd X}{P \ni X}$$

❑ If P believes the freshness of a formula X, then P is also of   (F1)
the view that each formula that has X as a partial formula is
fresh and that each function value $F(X)$ computable with a
feasible effort is fresh (F1, 'F' denotes 'freshness' rules on the
currentness of formulas):

$$\frac{P \mid\equiv \#(X)}{P \mid\equiv \#(X,Y), \ P \mid\equiv \#(F(X))}$$

❑ If P believes in the recognisability of a formula X, then it   (R1)
can be concluded that P also believes in the recognisability
of each combined formula containing X as a partial formula
and that P is convinced of the recognisability of a computationally feasible function $F(X)$ (R1, 'R' denotes 'recognisability' rules for formulas):

$$\frac{P \mid\equiv \phi(X)}{P \mid\equiv \phi(X,Y), \ P \mid\equiv \phi(F(X))}$$

❑ Let us say P receives a formula $\{X\}_K$ encrypted with key K (I1)
that it did not create itself and P possesses key K and considers it an appropriate key to secure communication with
Q. If P also believes that message X is recognisable and X
or K is current, then it can be interpreted that P believes Q
conveyed messages X and $\{X\}_K$ and possesses key K (I1, 'I'
denotes rules on the 'interpretation' of formulas).

$$\frac{P \lhd \star \{X\}_K, \ P \ni K, \ P \mid\equiv P \xleftrightarrow{K} Q, \ P \mid\equiv \phi(X), \ P \mid\equiv \#(X,K)}{P \mid\equiv Q \mid\sim X, \ P \mid\equiv Q \mid\sim \{X\}_K, \ P \mid\equiv Q \ni K}$$

❑ If P believes that Q has jurisdiction over statement C and if   (J1)
P holds the view that Q believes in the correctness of statement C, it can be deduced that P also believes in the correctness of statement C (J1, 'J' denotes 'jurisdiction' rules):

$$\frac{P \mid\equiv Q \mid\Longrightarrow C, \ P \mid\equiv Q \mid\equiv C}{P \mid\equiv C}$$

(J2)   ❏ If P holds the view that Q is honest and competent in respect of all its beliefs ($P \mid\equiv Q \mid\Longrightarrow Q \mid\equiv \star$) and if P holds the view that Q sent message X on the condition C and that message X is fresh, it can be concluded that P holds the view that Q believes in the correctness of statement C **(J2)**:

$$\frac{P \mid\equiv Q \mid\Longrightarrow Q \mid\equiv \star, \; P \mid\equiv Q \mid\sim (X \rightsquigarrow C), \; P \mid\equiv \#(X)}{P \mid\equiv Q \mid\equiv C}$$

The reader is referred to [GNY90] for a comprehensive introduction to GNY-Logic and a complete list of all its rules. The use of the logic below is clarified using an analysis of the Needham–Schroeder protocol.

7.7.3   An example of GNY-Logic

The Needham–Schroeder protocol, introduced in Section 7.3, will be examined again to illustrate analysis based on GNY-Logic. In summary it consists of the following messages:

Protocol Definition

1. $A \to TTP$: (A, B, r_A)

2. $TTP \to A$: $\{r_A, B, K_{A,B}, \{K_{A,B}, A\}_{K_{B,TTP}}\}_{K_{A,TTP}}$

3. $A \to B$: $\{K_{A,B}, A\}_{K_{B,TTP}}$

4. $B \to A$: $\{r_B\}_{K_{A,B}}$

5. $A \to B$: $\{r_B - 1\}_{K_{A,B}}$

A syntactical analysis of the protocol leads to the following formulas[1]:

Syntactical Analysis

1. $TTP \lhd \star A, \star B, \star r_A$

2. $A \lhd \star\{r_A, B, \star K_{A,B} \rightsquigarrow TTP \mid\equiv A \overset{K_{A,B}}{\longleftrightarrow} B\}_{K_{A,TTP}}$
   $\rightsquigarrow TTP \mid\equiv A \overset{K_{A,B}}{\longleftrightarrow} B$

3. $B \lhd \star\{\star K_{A,B}, \star A\}_{K_{B,TTP}} \rightsquigarrow TTP \mid\equiv A \overset{K_{A,B}}{\longleftrightarrow} B$

4. $A \lhd \star\{\star r_B\}_{K_{A,B}}$

5. $B \lhd \star\{\star F(r_B)\}_{K_{A,B}} \rightsquigarrow A \mid\equiv A \overset{K_{A,B}}{\longleftrightarrow} B$

---

[1]The only message parts of each entity that are listed are those of interest in this context. Therefore, the partial message encrypted with $K_{B,TTP}$ is not given for A in the second step because A cannot decrypt it and thus cannot make any further inferences from it.

A formalisation of the initial conditions leads to the following for-   *Conditions*
mulas:

❏ The two entities A and B possess their own authentication
key $K_{A,TTP}$ or $K_{B,TTP}$ and believe in the appropriateness of
this key:

$$A \ni K_{A,TTP}; \qquad A \mid\equiv A \overset{K_{A,TTP}}{\longleftrightarrow} TTP$$

$$B \ni K_{B,TTP}; \qquad B \mid\equiv B \overset{K_{B,TTP}}{\longleftrightarrow} TTP$$

Furthermore, A and B have random numbers and believe in
their freshness. A believes that the name of B is recognisable
and B believes that the random number r_B is recognisable:

$$A \ni r_A; \qquad A \mid\equiv \#(r_A); \qquad A \mid\equiv \phi(B)$$

$$B \ni r_B; \qquad B \mid\equiv \#(r_B); \qquad B \mid\equiv \phi(r_B)$$

Finally, both A and B believe in the honesty of trusted entity
TTP and that TTP can generate an appropriate session key
for secure communication between A and B:

$$A \mid\equiv TTP \mid\Longrightarrow TTP \mid\equiv \star; \qquad A \mid\equiv TTP \mid\Longrightarrow (A \overset{K_{A,B}}{\longleftrightarrow} B)$$

$$B \mid\equiv TTP \mid\Longrightarrow TTP \mid\equiv \star; \qquad B \mid\equiv TTP \mid\Longrightarrow (A \overset{K_{A,B}}{\longleftrightarrow} B)$$

❏ The trusted entity possesses the necessary keys $K_{A,TTP}$,
$K_{B,TTP}$ and $K_{A,B}$ and believes in their respective secretness:

$$TTP \ni K_{A,TTP}; \qquad TTP \mid\equiv A \overset{K_{A,TTP}}{\longleftrightarrow} TTP$$

$$TTP \ni K_{B,TTP}; \qquad TTP \mid\equiv B \overset{K_{B,TTP}}{\longleftrightarrow} TTP$$

$$TTP \ni K_{A,B}; \qquad TTP \mid\equiv A \overset{K_{A,B}}{\longleftrightarrow} B$$

The rules of GNY-Logic can be used to make the following inferences for each protocol run [GNY90]:

1. With rules T1 and P1 it can be deduced from *Message 1*   *TTP possesses*
   that the trusted entity knows formulas A, B and r_A: $TTP \ni$ *formulas A, B and r_A*
   A, B, r_A.

2. With *Message 2* it should be noted that extension $\rightsquigarrow TTP \mid\equiv$ *A possesses the*
   $A \overset{K_{A,B}}{\longleftrightarrow} B$ of the formulas is correct as it directly follows from   *session key and*
   the initial conditions. It can also be assumed that receiver A *believes in its secrecy*

of this message cannot mistakenly use key $K_{A,B}$ for a different communication partner than B as the name of entity B is contained in the message.

Through rules T1, T3 and P1 it can now be inferred that A knows the contents of the message: $A \ni (r_A, B, K_{A,B}, \{K_{A,B}, A\}_{K_{B,TTP}})$. In particular, using Rule T2, A knows the session key: $A \ni K_{A,B}$.

With Rule F1 it can also be inferred that A believes in the freshness of this message, i.e.,

$A \mid\equiv \#(r_A, B, K_{A,B}, \{K_{A,B}, A\}_{K_{B,TTP}})$.

Likewise, through Rule R1 it is inferred that A believes in the recognisability of this message, i.e.,

$A \mid\equiv \phi(r_A, B, K_{A,B}, \{K_{A,B}, A\}_{K_{B,TTP}})$.

Through Rule I1 it can now be concluded that A believes that trusted entity TTP created the message, thus

$A \mid\equiv TTP \mid\sim (r_A, B, K_{A,B}, \{K_{A,B}, A\}_{K_{B,TTP}})$.

Using Rule J2, we obtain that A believes that TTP considers the session key to be secret: $A \mid\equiv TTP \mid\equiv B \stackrel{K_{B,TTP}}{\longleftrightarrow} TTP$.

Therefore, through Rule J1 it can be inferred that A also believes the session key is secret: $A \mid\equiv B \stackrel{K_{B,TTP}}{\longleftrightarrow} TTP$.

B possesses the session key

3. Regarding *Message 3* it should be noted that the extension $\rightsquigarrow TTP \mid\equiv A \stackrel{K_{A,B}}{\longleftrightarrow} B$ is inferred from the conditions and is therefore correct. With rules T1, T3 and P1 it can be inferred that B possesses the session key, i.e., $B \ni K_{A,B}$.

At this juncture no other useful conclusions can be reached using the rules of GNY-Logic. The reason for this is that one cannot infer that B believes in the currentness of the message received from A. In fact, this message could also turn out to be a replay of a previously recorded message.

A possesses r_B

4. Through rules T1, T3 and P1, it can be concluded from *Message 4* that A knows random number r_B, thus $A \ni r_B$. However, other conclusions cannot be reached. The reason for this is that A is unable to recognise the contents of Message 4, because even without knowing key $K_{A,B}$, an attacker can send a ciphertext that when decrypted provides a bit pattern that could be interpreted by A as a random number. The attacker in this case would not know the random number assumed by A, but this would be unnecessary anyway at this point in the protocol.

5. No further useful inferences can be made from *Message 5* for achieving the objectives of the protocol. This is because B has been unable to arrive at any beliefs about the session key $K_{A,B}$.

B believes neither in the freshness nor in the secrecy of the session key

Overall this means that the protocol provides no useful new beliefs after the third message; therefore, messages 4 and 5 can just as well be eliminated without causing any change to the result of a protocol run. A possesses the session key and believes in its secrecy, whereas B only possesses the session key but associates no further beliefs with it.

The introduction to GNY-Logic [GNY90] mentioned earlier also explains how the approach can be used to prove how the intended protocol objectives for a modified version of the Needham–Schroeder protocol can be achieved. For reasons of space we will not present these explanations because our primary aim in this section is to provide an initial introduction to the formal analysis of cryptographic protocols. We highly recommend that interested readers study the cited article because it provides important background for dealing with such protocols.

7.8 Summary

Cryptographic protocols are defined through a series of processing steps and message exchanges between multiple entities with the purpose of achieving a specific security objective. Numerous different cryptographic protocols are available for implementing diverse applications, with authentication and key management representing the most important applications for network security.

Definition of cryptographic protocols

The cryptographic protocols for entity authentication can be divided into the two principal categories of *protocols with trusted entity* and *direct authentication protocols*.

Categories

Representatives of the first category include the insecure *Needham–Schroeder protocol*, the (secure) *Otway–Rees protocol* as well as the *Kerberos* authentication and authorisation system for workstation computer clusters.

Needham–Schroeder, Kerberos

The international Standard X.509 defines data formats and techniques for the certification of public keys and three modular direct authentication protocols that can be implemented either with asymmetric or symmetric cryptographic algorithms.

X.509

Because the definition of secure cryptographic protocols has proven to be error-prone in the past, formal validation of new

Formal validation

cryptographic protocols is highly advisable. A number of different approaches are available for this purpose, with logic-based approaches such as GNY-Logic proving to be particularly successful at detecting security flaws.

7.9　Supplemental Reading

[Bry88]　BRYANT, R.: *Designing an Authentication System: A Dialogue in Four Scenes.* 1988. – Project Athena, Massachusetts Institute of Technology, Cambridge, USA

[GNY90]　GONG, L.; NEEDHAM, R. M.; YAHALOM, R.: Reasoning About Belief in Cryptographic Protocols. In: *Symposium on Research in Security and Privacy* IEEE Computer Society, IEEE Computer Society Press, May 1990, pp. 234–248

[KNT94]　KOHL, J.; NEUMAN, B.; TS'O, T.: The Evolution of the Kerberos Authentication Service. In: BRAZIER, F.; JOHANSEN, D. (Eds): *Distributed Open Systems*, IEEE Computer Society Press, 1994

[NS78]　NEEDHAM, R. M.; SCHROEDER, M. D.: Using Encryption for Authentication in Large Networks of Computers. In: *Communications of the ACM* 21, December 1978, No. 12, pp. 993–999

[NS87]　NEEDHAM, R.; SCHROEDER, M.: Authentication Revisited. In: *Operating Systems Review* 21, 1987

[OR87]　OTWAY, D.; REES, O.: Efficient and Timely Mutual Authentication. In: *Operating Systems Review* 21, 1987

7.10　Questions

1.　Does it make sense to talk about message integrity without also considering data origin authentication? What is the opposite situation?

2.　Why is merely exchanging authentic messages not sufficient for implementing entity authentication?

3.　Must a trusted entity always be directly incorporated into a cryptographic protocol?

4. Name the typical tasks of a trusted entity and list as many examples of protocols as possible where a trusted entity is either directly or indirectly involved.

5. Explain why Kerberos makes a distinction between ticket-granting and service-granting tickets.

6. Why are certificates only valid for a limited period of time?

7. Can you execute an authentication test completely 'offline' on the basis of certified keys?

8. Why are two message exchanges not sufficient for mutual authentication without synchronised clocks?

9. Can the property of Perfect Forward Secrecy be guaranteed with session key exchange based on RSA encryption? Would this be possible with ElGamal encryption?

10. List the typical objectives of an authentication protocol with integrated key negotiation.

8 Access Control

Previous chapters have already dealt with a range of algorithms and schemes that all play a specific part in guaranteeing that data and resources in information-processing systems are protected against unauthorised eavesdropping, and are available to legitimate users. Important prerequisites include that the entities of a system can be clearly and securely identified, that data is protected against eavesdropping and modification by cryptographic schemes, and that events that have taken place can subsequently be proven to third parties.

This chapter expands on the issue of how decisions are made and implemented in a system in order to determine which entities are authorised access to certain services and information. We start by introducing some central terms and concepts. The section that follows provides an explanation of 'security labels' that are integrated in some classic multi-user operating systems. The next section addresses the specification of access control guidelines. The chapter concludes with a general overview of the basic categories of access control mechanisms.

8.1 Definition of Terms and Concepts

Definition 8.1 *The term* **access control** *describes the process of mediation between the requests of the* **subjects** *of a system to perform certain* **actions** *on specific* **objects**. *The main task of access control is to decide, based on a defined* **security policy**, *whether or not a specific access can be permitted and to enforce this decision.*

Access control

The concept of the *reference monitor*, illustrated in Figure 8.1, is a central model of access control. A reference monitor is an imaginary entity within a system. All access requests are directed to the reference monitor for verification, and based on the security policies defined in the system, it decides whether access should be permitted. In real systems, this concept is normally not implemented as a special system entity and the functionality of the reference

A reference monitor is only a conceptual entity

Figure 8.1
*The concept of a
reference monitor*

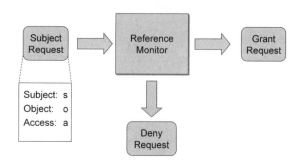

monitor is distributed over a number of system functions. It is, therefore, a conceptual rather than an actual entity.

Security policy **Definition 8.2** *The **security policy** of a system defines the conditions under which access requests by subjects that want to access specific objects by performing certain actions on them are mediated and enforced by the reference monitor functionality of a system.*

The term security policy is often also interpreted in a wider sense that encompasses all security aspects of a system including threats, risks, security objectives, possible countermeasures, etc.

Access control policy In this book we sometimes use the wording *access control policy* when we want to clarify that a security policy primarily refers to the definition of access rights.

The definitions introduced to this point are normally given in the context of the security of computers and operating systems. In this connection subjects and objects are defined as follows:

Subject, object **Definition 8.3** *A **subject** is an active entity that can initiate requests in order to access certain resources and then uses these resources to complete specific tasks. An **object** is a passive repository used to store information.*

Examples of objects based on this definition include files, directories, etc., whereas subjects are realised through processes in the sense of operating systems.

*Relationship to
communication
systems* However, in the context of communication systems it is not always easy to make a clean separation between subjects and objects. For instance, when an entity sends a message to another entity, it is not necessarily clear whether the receiving entity should be considered as an object or as a subject.

Type of access Furthermore, an understanding is needed in this context about what 'access' is in a communication system, and what types of access exist. In classic operating system tutorials, a distinction is made between the access types 'read', 'write' and 'execute,' whereas in the context of object-oriented programming each method offered by an object interface is defined as its own access type.

For example, an initial naïve approach would be to define the two access types 'send' and 'receive' for communication systems. As we will see in Chapter 13, intermediate systems only transmit data streams and so this distinction proves to be of little help when implementing an access control policy for networks.

8.2 Security Labels

We have seen that a direct transfer of access control concepts from operating systems to communication systems is not always obvious. Nevertheless, we will discuss these concepts further in this chapter because some of them have found acceptance, in one form or the other, in the implementation of access control in communication networks. This section therefore introduces the concept of *security labels*.

Definition 8.4 *A **security level** is defined as an hierarchical attribute that indicates the level of sensitivity of the subjects and objects of a system.*

Security level

The term 'hierarchical attribute' is often used to express that the level in each case can be brought into a strict order ('<'). An example of security levels in the military is the categorisation of information as 'unclassified' < 'confidential' < 'secret' < 'top secret'. The commercial sector tends to use the labels 'public' < 'sensitive' < 'proprietary' < 'restricted.' There is no major difference between them, as what is essential in both classifications is the strict order that exists regardless of the respective names given to the labels.

Definition 8.5 *A **security category** is a non-hierarchical grouping of subjects or objects that simplifies the process of indicating the level of their sensitivity.*

Security category

Security categories combine subjects or objects that are awarded the same access capabilities. This leads to the use of the 'Need to know' principle, which requires that an entity should only access the information it needs to complete its tasks. In the military, for example, a general in the army is not necessarily able to access

'Need to know' principle

confidential information from the air force, even if he is authorised to access information classified as 'top secret' for the armed forces. An example from the commercial sector would be the separation of an organisation's important information into diverse product areas and management.

Security label **Definition 8.6** *A **security label** is an attribute that is associated with the entities (subjects, objects) that exist in a system and indicates their hierarchical level of sensitivity and their security category. Mathematically this is expressed as follows:*

$$Labels = Levels \times \mathcal{P}(Categories)$$

Here $\mathcal{P}(S)$ refers to the *powerset* of the set S, i.e., the set of all subsets of S.

Binary relations on the set of security labels constitute an important concept for specifying security policies based on security labels. A binary relation on a quantity S is a subset of the cross *Dominates relation* product $S \times S$. An example is the *dominates relation* that enables a comparison between two security labels:

$$Dominates\colon Labels \times Labels$$
$$Dominates = \{(b_1, b_2) \mid b_1, b_2 \in Labels \land$$
$$level(b_1) \geq level(b_2) \land$$
$$categories(b_2) \subseteq categories(b_1)\}$$

This relation implies that a security label b_1 dominates another label b_2 if its level of sensitivity is greater than or equal to that of the dominated label and if the set of security categories of the dominated label b_2 is a subset of the set of the dominating label b_1.

8.3 Specification of Access Control Policies

Mathematical notation is often used in the specification of access control policies. Before introducing this notation, let us consider the following mappings:

$$allow\colon Subjects \times Accesses \times Objects \to Boolean$$
$$own\colon Subjects \times Objects \to Boolean$$
$$admin\colon Subjects \to Boolean$$
$$dominates\colon Labels \times Labels \to Boolean$$

These mappings always produce a Boolean output value, i.e., 'true' or 'false', depending on whether or not the statement of facts suggested by the name of the mapping has been fulfilled. Thus, for example, the 'allow' mapping produces the value 'true' for precisely all tuples (s, a, o) where the subject s has authorisation to use action a to access object o. The mapping 'own' enables us to express whether an object is owned by a specific subject, and 'admin' indicates whether a subject is a member of the set of system administrators. The mapping 'dominates' supplies the value 'true' precisely for all tuples (b_1, b_2) for which $(b_1, b_2) \in Dominates$ holds.

A number of general access control policies can be defined on the basis of these structures:

General access control policies

$$ownership\colon \forall\, s \in Subjects, o \in Objects, a \in Accesses\colon$$
$$allow(s, o, a) \Leftrightarrow own(s, o)$$

$$own\_admin\colon \forall\, s \in Subjects, o \in Objects, a \in Accesses\colon$$
$$allow(s, o, a) \Leftrightarrow own(s, o) \lor admin(s)$$

$$dom\colon \forall s \in Subjects, o \in Objects, a \in Accesses\colon$$
$$allow(s, o, a) \Leftrightarrow dominates(label(s), label(o))$$

The last *'dom' policy* listed requires that a system stores and processes security labels for the entities (subjects, objects) that exist in the system. Its advantage over other policies is that it enables complex access control policies to be implemented.

8.4 Categories of Access Control Mechanisms

Concrete mechanisms are required to implement the concepts described so far into real systems. This section presents a general classification of the proposed mechanisms.

Definition 8.7 *The concept of a reference monitor is implemented in real systems through an* **access control mechanism**.

Access control mechanism

The access control mechanisms proposed in the literature are normally divided into the following categories [SV01]:

❑ *Discretionary access control* comprises the procedures and mechanisms that mediate access requests at the discretion of

Discretionary access control

individual users. An example of this category of access control mechanisms is administration of access rights for files in the Unix operating system, where users can independently define the access rights for files created by them.

Mandatory access control

❏ *Mandatory access control* refers to the procedures and mechanisms that enforce the mediation of access requests by means of a uniform set of rules specified by a central entity.

Role-based access control

❏ *Role-based access control* is a more recent development that was particularly popular during the 1990s and which also underwent further development at that time. Conceptually, it supplements the two classic approaches listed above. The key innovation compared with classic approaches is its approach to access. It states that the decision as to whether a subject may use a specific method to access a certain object should not depend on the identity of the subject or on a general policy, but be determined by the concrete role in which the subject perceives his task. This simplifies the rights administration process [SCFY96].

What also applies is that the two classic approaches can be combined, with the mandatory access control decisions normally overriding the discretionary access of the user. An example is the use of discretionary access control on single computers combined with mandatory access control in dedicated intermediate systems of a communication network (also see Chapter 13 on Internet firewalls).

Access matrix

The *access matrix*, in which the rows in a system represent existing subjects and the columns the objects, has proven to be a useful concept for defining access control mechanisms [Amo94]. Each cell in the matrix contains the access rights of the subject addressed by the row to the object addressed by the column (also see Figure 8.2).

Implementation of access matrix

Because most entries in this conceptual matrix will be empty in real systems, direct implementation of the matrix is not very efficient in terms of storage consumption. Consequently, various other implementation designs have established themselves:

Access control lists

❏ For each object in a system, *access control lists (ACL)* store lists of the subjects that are permitted access to the object, sometimes in conjunction with the specific access rights of each individual subject. Access control lists are normally used together with discretionary access control because the

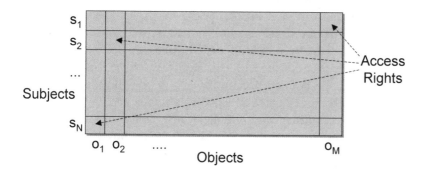

Figure 8.2
The conceptual access control matrix

quantity of ACLs makes them difficult to be maintained by a central administrative facility.

❑ To an extent, *capabilities* are a complementary concept to ACLs , because a list is stored for each subject containing the objects and the specific access types that the subject can use to access a specific object. The advantage, and possibly the danger, of capabilities lies in the fact that they make it viable for rights to be delegated between subjects. Consequently, they are normally used in conjunction with discretionary access control.

Capabilities

❑ *Security labels* are mostly combined with mandatory access control because the generation and administration of security labels can easily be automated through a centrally specified system function.

Security labels

What is common to all implementation alternatives for administration of access rights mentioned above is that the integrity of the respective data structures is an essential prerequisite for the security achievable by the scheme. If these data structures cannot be effectively protected from unauthorised manipulation by individual subjects, then all access control procedures are doomed to fail in their task of protecting objects from unauthorised access.

Integrity of access data structures is crucial!

8.5 Summary

Access control encompasses the administration and implementation of access rights that regulate under which conditions and with which actions the subjects of a system can access the objects of the system.

Tasks of access control

Allocation of rights   Depending on the access control mechanisms used, these rights can be allocated either on the basis of discretionary access control or mandatory access control. A simpler form of access rights administration involves a definition of roles where specific access rights are allocated in a second step. The rights for specific subjects can be defined on the basis of a possibly dynamic allocation of subjects to certain roles (role-based access control). In practice the fundamental approaches are often combined.

Reference to Internet firewalls   Even though there may not always be an obvious reason for transferring the fundamentals presented in this chapter to situations that exist in communication networks, they do represent central concepts for understanding access control issues in networks. This topic will be dealt with again in Chapter 13 on Internet firewalls.

8.6 Supplemental Reading

[Amo94]   AMOROSI, E. G.: *Fundamentals of Computer Security Technology*. Prentice Hall, 1994
Chapter 22 deals with access control mechanisms; other basic concepts related to access control are introduced in Chapters 6 to 13.

[SV01]   SAMARATI, P.; DE CAPITANI DI VIMERCATI, S.: Access Control: Policies, Models, and Mechanisms. In: FOCARDI, R.; GORRIERI, R. (Eds): *Foundations of Security Analysis and Design; Lecture Notes in Computer Science* Vol. 2171, Springer, 2001, pp. 137–196

[SCFY96] SANDHU, R.; COYNE, E.; FEINSTEIN, H.; YOUMAN, C.: Role-Based Access Control Models. In: *IEEE Computer* 29 February 1996, No. 2, pp. 38–47

8.7 Questions

1.   What does the term 'reference monitor' mean?

2.   How are the terms security level, security category and security label connected?

3.   Which of the access control policies *ownership, own_admin* or *dom* is normally used for the file system of the Unix operating system?

4. To which of the three categories of access control mechanisms – discretionary, mandatory or role-based – would you allocate rights assignments to files common in Unix?

5. Do access control lists support delegating rights between subjects?

Part II

Network Security

9 Integration of Security Services into Communication Architectures

The fundamentals of data security technology introduced in Part I of this book will be discussed in the following chapters from the perspective of their use in architectures and protocols of network security. The first chapter discusses basic issues and each of the three chapters that follows introduces and explains the security protocols of layers 2, 3 and 4. This part concludes with Chapter 13 on Internet firewalls, which implement access control in networks based on IP technology.

The first section of this chapter starts with the motivation for the basic design issues that are raised with the integration of security services into communication architectures. Section 9.2 presents a pragmatic model for secure networked systems. Using this model as the basis, the three remaining sections examine general and specific issues concerning placing security services in layered communication architectures.

9.1 Motivation

Analogous to the structured approach for conducting a security analysis for a given communication architecture (also see Section 1.3), the issues that arise in conjunction with the integration of security services into communication systems can also be categorised according to two dimensions. Along a horizontal orientation, there is the question of which security services should be implemented *into which systems* of a communication network (also see Figure 9.1). Individual security services can conceivably be provided in end systems or intermediate systems. Regarding integration into intermediate systems, a further differentiation can be made as to which security services should be embedded into which intermediate systems, for example, whether these services should be inte-

Two dimensions guide the categorisation of design decisions

Figure 9.1
*Fundamental design
decisions for network
security (1)*

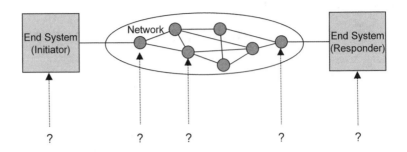

Dimension 1:
Which security service should be realized in which node?

grated solely on the 'boundary nodes' of a network or also generally within the network.

*Integration into
end systems or
intermediate systems?*

On one hand it could be argued that security services should basically be supplied in end systems because it is only there that users have total control over them and can be sure that their data will have the desired protection. However, on the other hand, the opposite approach can be taken with the argument that the user's control over security services is what actually leads to the source of insecurity (refer also to the principles of 'discretionary' and 'mandatory access control' in Chapter 8). However, as this chapter will show, the requirements that come to light and the constraints needing consideration are of a very diverse nature.

Figure 9.2
*Fundamental design
decisions for network
security (2)*

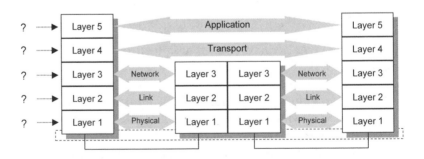

Dimension 2:
Which security service should be realized in which layer?

On the other hand, design decisions are also needed along a vertical orientation to identify *the layers* into which specific security services should be integrated (see Figure 9.2). Here again it is possible to take one or the other extreme position, dictating that security services should basically be supplied in the application layer because only applications have a complete knowledge of the semantics and thus the sensitivity of potentially protected data. It is just as easy to argue that basically all data, including the protocol information of the deeper layers, requires equal protection to ensure that the best possible security is achieved. This would mean that security services should be extended to the deepest protocol layer possible. Even if both arguments have a certain validity, the aspects of real networks concern so many layers that the issue of security cannot be resolved satisfactorily using a simple 'wholesale solution'.

Integration into which protocol layer?

'One size does not fit all!'

In this chapter we will take a close look at these aspects from a general perspective. The simple, pragmatic model of secure networked systems in the next section provides a useful orientation for our discussion.

9.2 A Pragmatic Model

Ultimately, the basic objective in setting up communication networks and distributed systems is to make certain applications available to the users of these systems. Based on this consideration, Figure 9.3 shows a pragmatic model for secure networked systems [For94].

In the model, a distinction is made between four principal *levels* where specific security requirements and measures can be embedded:

❑ The *application level* relates to requirements that have to be met for specific applications and the relevant measures that are provided directly in the applications themselves. Applications are software objects that handle certain tasks, such as electronic mail, WWW services, word processing, data storage, etc.

Application level

❑ The *end-system level* relates to requirements and measures that should be addressed uniformly between end systems. End systems are devices ranging from personal computers through servers to mainframes. The security policy of an end

End-system level

Figure 9.3
A pragmatic model for secure networked systems

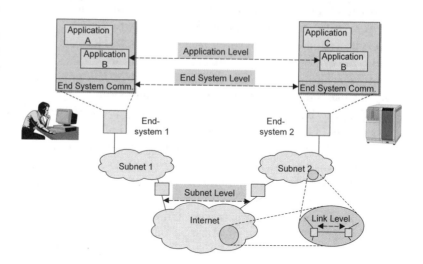

system is normally more or less uniformly specified by an entity, also called a *policy authority*, for the entire end system.

Subnetwork level

❏ At the *subnetwork level*, requirements and measures for communication facilities that fall under the control of an administrative organisation (e.g. department, company) should be dealt with on a uniform basis. The main idea is to ensure that communication between certain subnetworks over potentially untrustworthy networks (e.g. the Internet) is uniformly secure. The security policies for the subnetworks are normally specified uniformly for all systems in a subnetwork.

Link level

❏ The *link level* relates to the security between the separate nodes of a communication network that are directly linked over a physical medium ('link').

Delineating the terms internetwork and subnetwork

The other term *internetwork* contained in Figure 9.3 refers to a set of interconnected subnetworks. What distinguishes an internetwork from a subnetwork, which can actually also consist of a series of interconnected networks (e.g. LANs) and is guided by a security policy that has been allocated by a policy authority, is that no uniform security policy exists for an internetwork. Consequently, an internetwork is not normally categorised as a trusted network.

Comparing the layer model for communication systems commonly used today, we can see that the levels shown in Figure 9.3 cannot be aligned one-to-one with the protocol layers. As Figure 9.4 illustrates, the only direct association is from the application level to the application layer as the security services supplied in this level are implemented in the application layer or directly in the application itself.

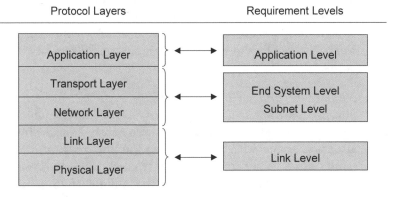

Protocol Layers   Requirement Levels

Figure 9.4
Mapping between protocol layers and requirement levels is not one-to-one

Security measures to meet the requirements of the end system and subnetwork levels can be implemented in the transport layer as well as in the network layer. Likewise, requirements at the link level can be implemented in the data link layer and the physical layer.

9.3   General Considerations for the Placement of Security Services

The following general considerations should be included in decisions on the placement of security services.

❑ *Mixing different data streams:* As a result of multiplexing in communication systems, there is a tendency in the lower levels or layers for data streams with different sources and sinks and from different applications to exist as mixed data streams. Normally, the security service for a specific level or layer will homogeneously handle the data traffic processed in this level or layer. However, this can lead to inadequate control of the security mechanisms being used on certain data. For example, if a security policy demands that traffic be handled specifically to deal with particular applications or users, this would be better implemented at a higher level.

Mixing data streams

Routing knowledge

❏ *Routing knowledge:* The lower levels usually have more knowledge about the security characteristics of different communication paths (e.g. routes or links). In environments where these characteristics vary significantly, placement of coordinated security mechanisms in the lower levels can provide considerable benefits in terms of effectiveness and efficiency. Communication paths that are particularly at risk (e.g. wireless links or routes traversing public networks between two interconnected local area networks) can therefore be specifically secured without those other parts of the network that are less at risk requiring additional security measures.

Number of protected points

❏ *Number of protected points:* When security services are placed on the application level, they must be implemented in every sensitive application and in each end system. A similar situation occurs with placement on the link level, as all end points must be secured from less trusted connection routes. Placement of security services in the 'middle' of an architecture tends to reduce the number of points needing protection.

Protection of protocol information

❏ *Protection of protocol information:* By nature the security measures of the higher protocol layers cannot protect the protocol fields of the layers below them. This is a point that should not be ignored, particularly as not only user data but also the network infrastructure itself has to be protected.

Source and sink binding

❏ *Source and sink binding:* Some security services, such as data origin authentication and non-repudiation, are based on a relationship between data and its sender and partly its receiver. This relationship can be established more effectively at the higher levels, particularly at the application level.

Beyond the general considerations listed, the following more level-specific observations also help to provide useful arguments for determining the most appropriate placement of individual security services for a specific network configuration.

Application level

In some cases, the *application level* is the only sensible level where certain security services can be implemented. An applications-specific security service, such as access control for a distributed file service, can only be implemented completely at the application level. It may also be necessary for a security service to be effective beyond certain application gateways. An example would be confidentiality and data origin authentication for electronic mail that is normally transported over multiple e-mail gateways before it is delivered to the receiver. Furthermore, the semantics of certain data elements may also require special security. For example, with the non-repudiation security service sufficient knowledge of the semantics of certain data elements exists only in the application itself. Lastly, the programmers of an application sometimes have no other choice but to integrate certain security services into the application level because the security mechanisms at the lower levels cannot be influenced by them.

The *end-system level* is particularly suitable when end sys- *End-system level* tems are categorised as trusted but the network in between is not trusted. Also, it is often advantageous if security services can be implemented transparently in relation to applications and if the configuration and management of security services can be transferred to a designated system administrator.

The *subnetwork level* should not be confused with the end- *Subnetwork level* system level, even if the security services are sometimes implemented in the same protocol layer. When security measures are implemented at the subnetwork level, the same protection is normally awarded to all end systems of the subnetwork concerned. The assumption is that a subnetwork connected directly to an end system is just as trustworthy as the end system itself. The basis for this assumption is that the end system and the subnetwork are sited at the same location and configured and administered by the same staff. The advantage of implementing security measures at the subnetwork level is that there are normally far fewer subnetwork gateways than there are end systems, and security measures are therefore needed for fewer systems.

The *link level* is especially recommended for the implemen- *Link level* tation of security measures when relatively few untrustworthy communication links exist. It is therefore easier and more cost-effective to secure only those links that are categorised to be insecure. Depending on the underlying technology, the link level also enables the use of specific protection mechanisms such as spread spectrum) transmission or key-dependent switching between different transmission frequencies (frequency hopping). The link level is also the only level where protection from traffic flow analysis can be realised effectively.

Another aspect in connection with security measures is the po- *Interaction with the* tential user interaction that cannot elegantly be integrated into *user* the current model, because users are outside the communication systems. For example, authentication in particular requires inter-action with a user. The following design alternatives are available:

Local authentication

❏ *Local authentication:* In this case, the users authenticate themselves to an end system. The end system in turn authenticates itself to the remote system and at the same time names the identity of the user. With this version, the remote end system must have confidence that the local end system correctly handled the authentication verification.

Specific protocol elements

❏ *Use of specific protocol elements at the application level:* A user supplies the local system with certain authentication information that the local system securely forwards to the remote system using specific protocol elements. In this case the remote system verifies the authentication itself.

Combined techniques

❏ *A combination of these techniques:* The Kerberos system presented in Section 7.4 can be used as an example that combines local and remote authentication as it undertakes both local and remote authentication.

9.4   Integration in Lower Protocol Layers vs Applications

This section supplements the discussion with additional considerations that support the integration of security services particularly, in the lower protocol layers:

Security

❏ *Security:* This can motivate the integration of security measures in lower protocol layers in two ways. Firstly, the network infrastructure itself must be protected so that it can guarantee its availability and ensure that it is functioning correctly, including its ability to provide an accurate and verifiable accounting of service usage. Secondly, security measures implemented in network elements are often more difficult for network users to attack that those in end systems. This is particularly the case when measures require hardware support for their implementation and cannot be deactivated or bypassed.

Application independence

❏ *Application independence:* Fundamental security services can be implemented once in the lower layers for all applications and do not have to be integrated into each individual application.

❏ *Quality of service:* Data streams that make special demands on the quality of a transmission service (minimal delay, delay fluctuation, etc.) can profit from being integrated into the lower protocol layers, where it is easier to combine the quality of service-orientated scheduling of a communication system with the scheduling of cryptographic operations. Take an example in which multiple data streams with different requirements and traffic characteristics (e.g. voice transmission and file transfer) are encoded by a hardware encryption module and then transmitted. In this case, the quality of service-orientated scheduling of both operations can in-tegrated more effectively if encryption is executed directly on the communication adapter before transmission. *Asynchronous Transfer Mode (ATM)*, which is the basis of *Broadband ISDN*, is an example of a communication architecture that can support such an integration, even securing end system to end system [ATM97a, ATM97b, ESSS98, Sch98, SHB95, Sta95, Sta98, SR97].

Quality of service

❏ *Efficiency:* Due to the improved possibilities for integrating hardware support directly onto a communications adapter, cryptographic operations can be executed more efficiently when integration is in the lower protocol layers. So although an application can basically use a special hardware extension to carry out an efficient computation of cryptographic oper-ations by calling up the appropriate functions, it has to be considered that this involves additional data transport over the system bus, which inevitably affects performance.

Efficiency

9.5 Integration into End Systems or Intermediate Systems

Aside from identifying the layer where certain security services are to be implemented, it is also important to establish whether particular security services should be integrated into end systems or into intermediate systems.

In respect to integration into end systems, it should be noted that this can generally take place at the applications level as well as at the end system level. In some circumstances it can make sense to incorporate security also at the link level — for instance, when a modem is used to connect an end system to a dedicated sys-tem. This would be the case if remote maintenance access exists for a switching system.

Integration into end systems usually addresses requirements of the applications or end system level

Integration into intermediate systems sooner tends to address requirements of the subnetwork or link level for user data

Integration into intermediate systems can take place at all four levels. Depending on the level, there is a tendency, however, for different security objectives to be pursued. Therefore, if security is integrated at the applications or end-system level, it tends to secure management interfaces of the intermediate systems rather than user data. The latter is more apt to occur at the subnetwork or link level.

Depending on the intended security objectives, integration into either end systems or intermediate systems can make sense. In practice both forms are often found.

Figure 9.5
Authentication relationships in internetworks

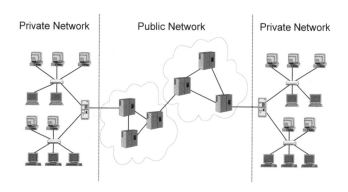

Authentication Relation			Application for securing
End System	↔	End System	User Channels
End System	↔	Intermediate System	Management Interfaces Accounting
Intermediate	↔	Intermediate System	Network Operation: Signalling, Routing, Accounting, ...

Authentication relationships in internetworks

Figure 9.5 presents an overview of authentication relationships in internetworks [Sch98]. As can be gleaned from this figure, authentication relationships in internetworks are needed between arbitrary combinations of different end and intermediate systems while at the same time pursuing completely different security objectives. Authentication between end systems (or the applications run in them) generally aims to secure actual user data. Authentication between end and intermediate systems, on the other hand, is aimed more towards securing management interfaces and a producing a verifiable accounting of service usage. Authentication between intermediate systems is mainly designed to secure network operations, securing signalling, the exchange of routing information and, in some cases, provides an important basis for the verifiable accounting of data transmitted between network gateways.

9.6 Summary

The integration of security services and mechanisms into communication architectures is guided by two main considerations, namely which security services should be implemented in which nodes and which security services should be realised in which layer.

A simple pragmatic model for secured networked systems that distinguishes between four different levels with different security requirements and where security measures can be taken is useful for design decisions.

Pragmatic model

As a number of arguments exist for and against each of these levels, there is no single correct solution to this design problem. A wholesale solution would not be sensible anyway, as different solutions have to be found based on the respective security objectives being pursued and this also means that different tradeoffs have to be taken into account. The security measures that result often present a compromise in terms of achievable security, performance, flexibility, etc.

A 'wholesale solution' does not exist

The remaining chapters in this part of the book closely examine numerous examples of network security protocols and architectures, thus enabling us to learn how to make a better assessment of the consequences of respective design decisions.

9.7 Supplemental Reading

[For94] FORD, Warwick: *Computer Communications Security – Principles, Standard Protocols and Techniques.* Prentice Hall, 1994
Chapter 3 introduces the integration of security measures into communication architectures and also describes the pragmatic model presented in this chapter.

9.8 Questions

The intention of the questions below is to integrate the knowledge presented in the whole of Part II. We therefore recommend that you continue with the following chapters and read this chapter again after Chapter 13 before responding to the questions.

1. Name the security objectives that particularly require that measures be integrated in nodes on the 'boundary' of a network.

2. Name a security protocol treated in this book that can be used either at the end system or at the subnetwork level.

3. Which of the security protocols discussed are particularly appropriate for exploiting knowledge about the security characteristics of certain data paths?

4. List the attacks from which a network infrastructure should be protected in particular, and the security objectives that have to be ensured in this context.

5. Assume that you could configure an e-mail gateway so that e-mails exchanged between two company locations are basically signed and encrypted between the two e-mail gateways. To which of the levels described would you allocate such a security measure?

6. Explain the advantages and disadvantages of the strategy presented in the previous question and discuss alternative concepts.

7. Which security protocols discussed in this book can implement security measures at the end system level?

10 Link Layer Security Protocols

According to the classic OSI model for open communication systems, the link layer (the second layer in the model) provides a reliable data transmission service between two peer entities that are directly connected over a shared medium (e.g. cable or radio transmission link). Its main tasks thus encompass *error detection and correction* and the control of joint access to shared mediums (e.g. Ethernet). The latter task is also referred to as *Medium Access Control (MAC)*. However, this term should not be confused with the description of access control given in Chapter 8. Likewise, use of the abbreviation MAC should make a distinction between its double meanings Medium Access Control and Message Authentication Code.

OSI Layer 2 tasks

Not all current communication protocols and technologies fit nicely into this model. For instance, dial-up connections to an Internet Service Provider (ISP) have characteristics that extend beyond those of the link layer, and some protocols for realising *virtual private networks (VPN)* perform tasks other than those assigned by the OSI model to Layer 2.

Fitting protocols into the OSI model is not always easy

As a result, an exact understanding of the term *security protocols of the link layer*, which interprets these as security protocols of the second OSI layer, does not appropriately reflect the reality of current communication architectures. In this book we therefore settle for the following more intuitive definition:

Definition 10.1 *The objective of a* **link layer security protocol** *consists of guaranteeing certain security properties of the protocol data units of the link layer, which means the protocol data units that transport the protocol data units of the network layer, e.g. IP packets.*

The following section presents an approach for securing access to local area networks that are based on the standards of the IEEE 802 series. This is followed by a definition of the point-to-point protocol, which is the protocol most frequently used over dial-up connections for access to the Internet and incorporates some fundamental security functions. Building on this discussion, Section 10.3 presents the point-to-point tunnelling protocol that enables the usage of PPP to be extended over the entire Internet. Section 10.4 concludes the chapter with comments on setting up virtual private networks.

10.1 Securing a Local Network Infrastructure Using IEEE 802.1x

IEEE-802 protocol suite

The *802-LAN/MAN-Standardisation Committee* of the *Institute of Electrical and Electronics Engineers (IEEE)* develops standards for Local Area Networks (LAN) and Metropolitan Area Networks (MAN). The most widely used standards are those of the Ethernet suite (IEEE 802.3), the Token-Ring (802.5) and the 802.11 series standards for wireless LANs (WLAN).

For the security of access to such networks a subgroup of the 802 committee is working on a standard labelled *IEEE 802.1x* [IEE01] that aims to restrict access to LAN services to those users or devices with proper authorisation. This standard can basically be used with diverse technologies of the 802 series.

Port-based access control

The basic characteristic of the standard is *port-based access control*, which is used to perform authentication and authorisation of devices connected to LAN ports. A LAN port is a logical (sometimes also physical) access point with point-to-point connection characteristics. It could be the access port of a Fast-Ethernet switch or the logical access point of a WLAN base station. The IEEE 802.1x standard conceptually distinguishes between two logical ports (also see Figure 10.1): an *uncontrolled port*, which enables a device to prove its identity through an authentication exchange, and a *controlled port*, which allows proven authenticated devices to access the general data transmission service of the local area network.

Roles of IEEE 802.1x

Three principal roles are distinguished in the authenticity verification of connected devices:

Supplicant

❑ A device that wants access to the data transmission service of the local area network finds itself in the role of *supplicant* when it is providing and proving its identity during the authentication exchange.

Authenticator

❑ The access point of the LAN infrastructure, such as an

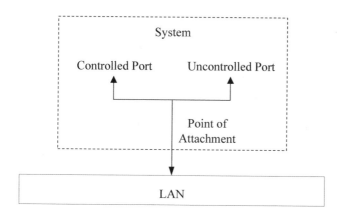

Figure 10.1
Controlled and uncontrolled ports with IEEE 802.1x

Ethernet-Switch, functions as an *authenticator* demanding that a device provide and prove its identity.

❑ The authenticator does not itself verify the credentials provided by a supplicant during the authentication exchange.
Instead, it forwards them to an *authentication server* that
then notifies it of the results of the authentication verification.

Authentication server

Prior to a device's successful authentication of itself to the authenticator of a local area network, it only has access to an uncontrolled port. This port is uncontrolled in the sense that it can
be accessed even before authentication has been successfully performed. However, it only allows authentication message exchange
and cannot be used for the transmission of arbitrary data units.

Access prior to authentication is only possible to the uncontrolled port

An authentication exchange can be initiated by a supplicant as
well as by an authenticator. The controlled port is opened as soon
as the exchange is successfully completed.

The IEEE 802.1x standard does not define any authentication protocols of its own but instead recommends the use of existing protocols, such as the *Extensible Authentication Protocol
(EAP)* [BV98] for basic authentication without key exchange or
the *PPP EAP TLS Authentication Protocol* standardised in RFC
2716, which also enables session keys to be negotiated during an
authentication exchange. IEEE 802.1x also recommends that the
authentication server is implemented in accordance with the IETF
specification for a *Remote Authentication Dial In User Service (RA-
DIUS)* [RWRS00].

IEEE 802.1x does not define its own authentication protocols

Figure 10.2
Protocol run of EAPOL protocol

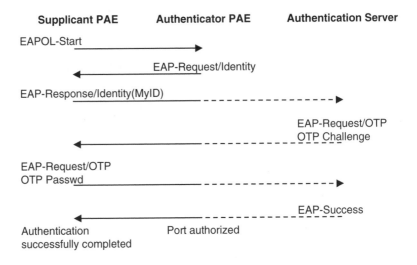

Supplicant PAE **Authenticator PAE** **Authentication Server**

EAPOL-Start

EAP-Request/Identity

EAP-Response/Identity(MyID)

EAP-Request/OTP
OTP Challenge

EAP-Request/OTP
OTP Passwd

EAP-Success

Authentication   Port authorized
successfully completed

EAP over LANs (EAPOL)

For the exchange of EAP protocol data units IEEE 802.1x specifies the protocol *EAP over LANs (EAPOL)*, which mainly defines techniques for the encapsulation of EAP-PDUs into the payload of transmission frames of the 802 protocol suite. The encapsulated PDUs are then exchanged between the *Port Access Entities (PAE)* of the supplicant and the authenticator. Conventional RADIUS messages can be used between the authenticator and the authentication server. Figure 10.2 presents an overview of the sample authentication of a device. Details of the EAP authentication protocol are given in Section 10.2.2 as EAP was specified in the context of the point-to-point tunnelling protocol.

IEEE 802.1x provides access control for LANs

In summary it should be noted that IEEE 802.1x primarily provides access control for the transmission services offered by local area networks. However, the standard does not define how to secure actual data transmission from passive or active attacks and additional security protocols are therefore required.

10.2 Point-to-Point Protocol

Large parts of the Internet are implemented through point-to-point connections. Examples include connections between Internet routers established over wide area networks, and dial-up connections used by hosts to gain access to an Internet service provider over the telephone network.

The two key protocols in this context are:

❑ *Serial Line IP (SLIP)*, which enables IP packets to be trans-   *Serial Line IP (SLIP)*
mitted over serial lines but neither performs its own error
detection nor supports dynamic address allocation, the transport of other network layer protocols or peer entity authentication [Rom88].

❑ *Point-to-Point Protocol (PPP)*, which was specified as a suc-   *Point-to-Point*
cessor to SLIP to deal with the inadequacies of SLIP men-   *Protocol (PPP)*
tioned [Sim94b, Sim94c].

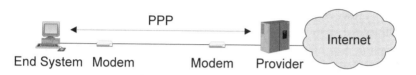

Figure 10.3
*Typical usage
scenario for PPP*

Figure 10.3 shows a typical usage scenario for PPP in which a computer using a modem over a telephone connection dials up a network provider that functions as a gateway to the Internet.

10.2.1 Structure and Frame Formats

The main components of PPP are: *PPP components*

❑ Layer 2 frame format for frame delineation and error detection;

❑ *Link Control Protocol (LCP)* the main task of which is connec-   *Link Control Protocol*
tion control (thus connection setup, test and release as well   *(LCP)*
as parameter negotiation);

❑ a range of *Network Control Protocols (NCP)* that are specif-   *Network Control*
ically tailored to the needs of the respective network layer   *Protocols (NCP)*
protocols to be transported in PPP frames (IP, IPX, NetBEUI,
Appletalk, etc.).

Figure 10.4 shows the Layer 2 frame format of PPP. The pro-   *Layer 2 frame format*
tocol uses character-oriented transmission, i.e., the transmission
frames are aligned to octet boundaries and character stuffing
is used to achieve code transparency.   With respect to layer
2 functions PPP basically realises the popular HDLC protocol.
Transmission frames therefore always begin with the bit pattern
'01111110', followed by the address and the control fields. The protocol field indicates which frame type is being transported in the

payload. For error detection a CRC checksum is computed and transmitted in the checksum field. The frame also closes with the characteristic bit pattern '01111110'.

Figure 10.4
Frame format of PPP

1	1	1	1 or 2	variable	2 or 4	1	Octets
Flag 01111110	Address 11111111	Control 00000001	Protocol	Payload	Checksum	Flag 01111110	

As Figure10.4 shows, unnumbered frames are usually transmitted. However, in usage scenarios with high error probability, a reliable transmission mode with sequence numbers and retransmission can also be negotiated. Unless a different size has been negotiated, the maximum payload size is 1500 octets.

Usage scenario 'Internet access'
A typical usage scenario for PPP is a personal computer (PC) accessing the Internet via a modem and dial-up connection over the telephone network:

❑ The modem of the user dials the telephone number of an Internet service provider (ISP) and sets up a 'physical' connection to the access computer of the ISP over the telephone network.

❑ The calling PC sends multiple LCP packets, which are always transmitted in PPP Layer 2 frames, in order to negotiate the desired PPP parameters.

❑ Optionally, a security-specific negotiation subsequently takes place (see below).

❑ The network layer is then configured through an exchange of NCP packets. In this connection the *Dynamic Host Configuration Protocol (DHCP)* optionally arranges for the dynamic allocation of an IP address.

❑ Once the connection is completely configured, the actual usage phase begins, during which arbitrary IP packets can be exchanged between the PC and any other computers in the Internet. The IP packets are transmitted in PPP packets from the PC to the access gateway that decapsulates and sends them to the next Internet router. Similarly, the IP packets destined for the PC from the access gateway are encapsulated into PPP packets and transmitted to the PC.

❏ During the connection termination phase the IP address reserved for the PC is released.

❏ Lastly, the Layer 2 connection is terminated using the LCP protocol and the modem closes down the 'physical' connection to the ISP access computer.

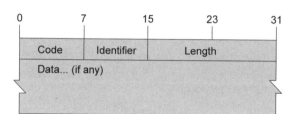

Figure 10.5
Frame format of PPP link control protocol

Figure 10.5 shows the frame format for the *Link Control Protocol (LCP)*. The significance of the fields contained is as follows:

LCP frame format

❏ The *code field* indicates the requested service primitive. The key primitives are called: Configure-Request, Configure-Ack, Configure-Nack and Configure-Reject for the request or confirmation of configuration commands; Terminate-Request and -Reject for the termination of a connection; Code-Reject and Protocol-Reject for the rejection of non-supported options; Echo-Request and -Reply and Discard-Request.

❏ The *identifier field* is used to map replies to previous requests.

❏ The *length field* indicates the length of the LCP packet including the LCP header (code, etc.).

❏ The *data field* is optional and contains command-specific information.

The configure primitives enable the agreement of Layer 2-specific configuration details, such as the maximum size of the payload field in Layer 2 frames, data compression and peer entity authentication.

The PPP specification recommends the optional execution of an authentication exchange after the setup of a Layer 2 connection. One of the two peer entities requests this exchange by signalling

PPP security services

the appropriate configure-request commands. Furthermore, encryption of user data can be negotiated after successful authentication. The respective security protocols are explained in detail in the sections below.

10.2.2 PPP Authentication Protocols

Two authentication protocols were defined for the first version of PPP [LS92]: *Password Authentication Protocol (PAP)* and *Challenge Handshake Authentication Protocol (CHAP)*. As time went by these protocols were supplemented by an extensible protocol called *Extensible Authentication Protocol (EAP)* [BV98], which was then extended by the version *EAP Transport Level Security Protocol (PPP-EAP-TLS)* [AS99].

Password authentication protocol
    The password authentication protocol (PAP) was defined in RFC 1334 [LS92] in 1992. It is a very simple protocol with the prerequisite that the authenticator knows a password of the peer entity. After successful setup of a Layer 2 connection, the authenticator requests that the peer entity use PAP to authenticate itself. The latter entity then sends an authenticate-request packet with its identity and a password. The authenticator verifies whether it knows the identity provided and whether the password is correct. If verification is successful, it replies with an authenticate-ack message whereas, if the authentication is faulty, it sends an authenticate-nack message. The protocol is insecure because it does not provide for any cryptographic protection of the password. It is therefore no longer mentioned in the more recent RFCs for PPP authentication [Sim96].

Challenge Handshake Authentication Protocol
    The Challenge Handshake Authentication Protocol (CHAP) was also specified in RFC 1334. It is a simple challenge-response protocol. Both peer entities must know a shared secret (thus an authentication key) they can use to perform the following authentication exchange [Sim96]:

❏ After the setup of a Layer 2 connection, one of the two entities (A) sends a challenge that consists of an identification *identifier*, a random number r_A and the name of A:

$$A \rightarrow B: \ (1, identifier, r_A, A)$$

❏ The peer entity B computes the value of a cryptographic hash function over its name, the shared secret $K_{A,B}$ and the random number r_A. It sends this value along with the identification of the protocol run and its name to A:

$$B \rightarrow A: \quad (2, identifier, H(B, K_{A,B}, r_A), B)$$

❑ Upon receipt of this message, A itself computes the hash value and compares it with the value received. If both values match, A responds with a success message.

The updated version of the protocol specification, RFC 1994, specifies that MD5 must be supported as a cryptographic hash function and that the use of any other hash functions can be negotiated.

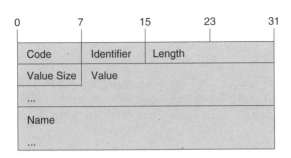

Figure 10.6
Frame format of PPP Challenge Handshake Protocol (1)

Figure 10.6 presents the frame format of the CHAP messages *Challenge* (Code 1) and *Response* (Code 2). The identifier is an octet that has to be changed with each challenge and is used to enable a simple mapping of responses to challenges. The length field indicates the total length of a message and 'value size' gives the length of the contained value (random number r_A or hash value according to description above). The last field, the name field, contains the identity of the sending entity. Its length is not explicitly transmitted and instead is computed by subtracting the length of the other fields from the total message length.

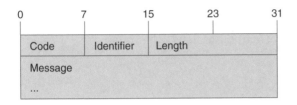

Figure 10.7
Frame format of PPP Challenge Handshake Protocol (2)

Figure 10.7 presents the format of the CHAP messages *Success* (Code 3) and *Failure* (Code 4). Compared with the messages explained above, these messages only contain an optional message element to indicate status messages or error situations. These messages are implementation dependent and are not interpreted by the protocol entities but only indicated to the user.

Extensible Authentication Protocol

The *Extensible Authentication Protocol (EAP)* is a general protocol that supports a range of diverse authentication methods that can also be more complex in nature than just 'a challenge plus a response.'

The protocol provides a series of basic commands:

❑ *Request, Response* are supplemented by a type field and type-specific data fields.

❑ *Success, Failure* are used to indicate the results of an authentication exchange.

Examples of type fields are *Identity, Notify, Nak* (only used in responses to indicate non-supported requests), *MD5 Challenge* (same as CHAP), *One-Time Password, Generic Token Card* and *EAP-TLS*.

One-time password

The basic idea behind the *'One-Time Password'* protocol *(OTP)* consists of transmitting a 'password' that can only be used for one run. This prevents a potential attacker who is eavesdropping on such a password from using it in future authentication attempts [HMNS98]. Prior to the initial protocol run, both peer entities A and B must perform an initialisation and agree on a shared starting value. The authenticator A chooses a random number r_A and sends it to supplicant B, who applies a hash function n times to compute a hash value over the random number and a personal secret value $Password_B$. The resulting value is the initial 'one-time' password of B: $PW_n = H^n(r_A, Password_B)$. The tuple (n, PW_n) must be transmitted to A in a 'secure' way (= confidentially and authentically) and stored there.

OTP protocol exchange

The actual authentication exchange then consists of the following steps:

1. The authenticator sends the number $n - 1$ to B:

$$A \to B: \quad n - 1$$

2. B then computes a new one-time password $PW_{n-1} := H^{n-1}(r_A, Password_B)$ and sends it to A:

$$B \to A: \quad PW_{n-1}$$

3. The authenticator now computes $H(PW_{n-1})$ and compares the result with PW_n. If both values match, A assumes that B is authentic and stores the tuple $(n-1, PW_{n-1})$ for the next authentication process. This exchange can be executed $(n-1)$ times before the procedure needs to be initialised again.

The security of this procedure is based on the fact that an attacker who eavesdrops on one of the one-time passwords PW_i will not be able to compute the value $H^{-1}(PW_i)$ that is necessary for the next authentication exchange.

The 'generic token card' method performs a simple challenge-response exchange. However, unlike CHAP, the user is also included in the exchange in which the random number selected by the authenticator is indicated to him or her. To compute the response, the user uses a small device the size of a credit card, called a *generic token card*. The user enters the random number onto the card and the card computes the appropriate response which the user then has to transfer manually to the computer. The advantages of this method are that users have better control over when they actually perform authentication and the card is more manipulation safe than a personal computer, as it does not have its own network connectivity and no complex software is installed on it. *Generic token card*

The *PPP-EAP-TLS* method uses the authentication exchange of the *Transport Layer Security* protocol specified in RFC 2246 [DA99]. As the TLS protocol operates in or above the transport layer, it is explained in Chapter 12. *PPP-EAP-TLS*

10.2.3 PPP Encryption

Once a PPP connection has been successfully authenticated, the encryption of the transmitted user data can be negotiated. The *Encryption Control Protocol (ECP)* [Mey96] defined in RFC 1968 is used to negotiate the method used and its parameters. ECP uses the same frame format as the link control protocol (LCP) and introduces two additional service primitives: *Reset-Request* and *Reset-Ack*. With the help of these primitives decryption errors can be indicated and acknowledged allowing for cryptographic resynchronisation (also see Section 2.5). *Encryption Control Protocol (ECP)*

The configure primitive is used to negotiate a specific cryptographic scheme with the desired scheme defined as the parameter. The values *DESE* and *3DESE* are predefined for simple or triple DES encryption and *proprietary* for the negotiation of proprietary schemes. Proprietary schemes are uniquely identified through a registered *Organizational Unit Identifier (OUI)* that identifies the vendor plus a vendor-specific value for the identification of a specified scheme.

A PPP frame can transport exactly one ECP packet

Exactly one ECP packet can be transported in the PPP data field. Two values are defined for identifying ECP packets in the PPP protocol field: '0x8053' for standard mode and '0x8055' for the separate encryption of individual link layer connections when several such connections exist to the same destination (e.g. with the simultaneous use of multiple ISDN connections to the same peer entity).

Figure 10.8
Frame format of PPP Encryption Control Protocol

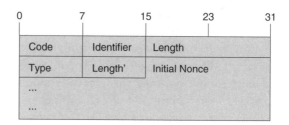

Figure 10.8 illustrates the frame format of an ECP message requesting a DESE encryption protocol. Similar to LCP, the request is made using a configure-request primitive (Code 1). The *type* field contains the value 3 when it requests DESEv2 (DESEv2 is an updated version which replaced the first version of the protocol). The *length'* field that follows indicates the total length of this configuration option (10 with DESEv2). The parameter *Initial Nonce* contains an eight octet long initialisation vector for DES in CBC mode (also see Section 3.1).

DESEv2

The format of PPP packets encrypted in accordance with the DESEv2 protocol [SM98b] is illustrated in Figure 10.9 and includes the link layer header for the encapsulation into HDLC frames. The address field contains the value '11111111' and the control field the value '00000011'. The value '0x0053' in the *protocol ID* field identifies the DESE protocol. The sequence number field that follows is initially assigned the value 0 and incremented by 1 with each packet transmitted by the sending peer entity. The length of the actual user data is padded to an integer multiple of 8 octets prior to encryption and then encrypted with the DES algorithm in CBC mode.

3DESE

The 3DESE protocol, which can be negotiated with the ECP protocol per configure-request with the type field set at value '0x02'

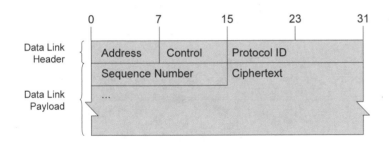

Figure 10.9
Format of encrypted PPP packets (DESEv2)

and is very similar to the DESEv2 protocol, can be used as an alternative to DESEv2 [Kum98]. The only difference between the two protocols is that 3DESE involves triple DES encryption with three different keys.

All encryption protocols for PPP are based on the assumption that an appropriate session key has been negotiated prior to encryption. This assumption is justified as the best time for the negotiation of a session key is during the authentication phase of a PPP connection. Such a negotiation is however only supported by one PPP authentication protocol, the PPP-EAP-TLS protocol.

Session keys are a prerequisite of PPP encryption protocols

10.3 Point-to-Point Tunnelling Protocol

The PPP protocol was originally specified as the successor to the Serial Line IP protocol and therefore only designed to be run directly between peer entities sharing a Layer 2 connection. An example is a PC that is connected over a telephone connection to the access computer of an Internet service provider. The basic idea behind the design of the *Point-to-Point Tunneling Protocol (PPTP)* was to extend the reach of PPP over the entire Internet by defining a method for transporting PPP packets as the payload of IP packets. This allows for PPTP being able to establish 'logical Layer 2 connectivity' over which arbitrary Layer 3 protocols such as IP, IPX, NetBEUI, Appletalk, etc., can operate.

Basic idea of PPTP

The payload of PPTP packets is therefore PPP packets without Layer 2-specific protocol fields such as HDLC information fields, the insertion of additional characters to achieve code transparency, CRC checksums, etc. For transport in the Internet the PPP packets are encapsulated into the payload of packets of the *Generic Routing Encapsulation Protocol (GRE)* that are then encapsulated

Content and transport of PPTP packets

into the payload of IP packets. Figure 10.10 shows the structure of the resulting PPTP packets.

Figure 10.10
Structure of PPTP packets

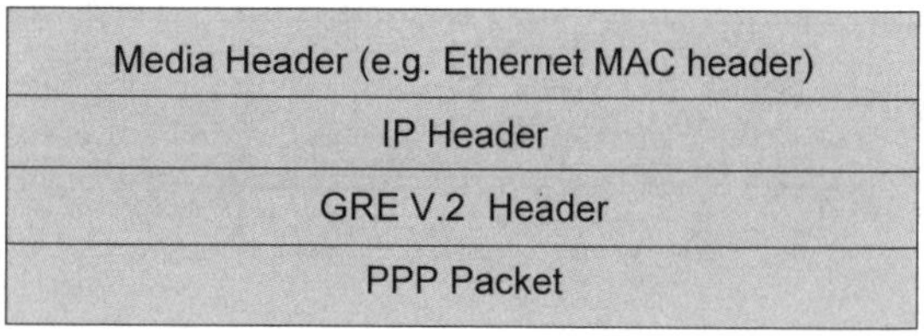

Media Header (e.g. Ethernet MAC header)
IP Header
GRE V.2 Header
PPP Packet

10.3.1   Basic Versions of PPTP Packet Encapsulation

Based on the encapsulation described, PPTP in a way creates a *tunnel* through the Internet in which PPP packets are transported. Such a tunnel can be realised between diverse entities with the following two types highlighting the kinds of differences that exist:

Voluntary tunneling

❏ Between a user PC and a *PPTP remote access server (RAS)*: With this version the user PC encapsulates the PPP packets itself and sends the resulting IP packets to the RAS server of the network where logical Layer 2 connectivity should be available. As the encapsulation takes place with the active involvement of the user PC, this version is also called *voluntary tunnelling*. The RAS server acts as the tunnel end point for the user PC in the destination network.

Compulsory tunnelling

The dial-up node functions as a proxy

❏ Between the local access number *(Point of Presence, POP)* of an ISP and a PPTP remote access server: This version does not involve the user PC in the decision about whether packets with PPTP should be tunnelled. It is therefore also called *compulsory tunnelling*. In this instance, the dial-up node of the ISP functions as a *proxy client* of the user PC *vis-à-vis* the RAS server. Although security can be realised at the subnetwork level with this version, true end-to-end security between the user PC and the RAS service cannot be achieved this way because the dial-up node of the ISP is the cryptographic peer entity of the RAS server. Consequently, for security reasons preference should be given to voluntary encapsulation.

Compulsory tunnelling

Figure 10.11 shows the frame structure on the different network segments when a user PC (client) accesses an application server that is located in an Intranet to which the client is attached through compulsory PPTP tunnelling.

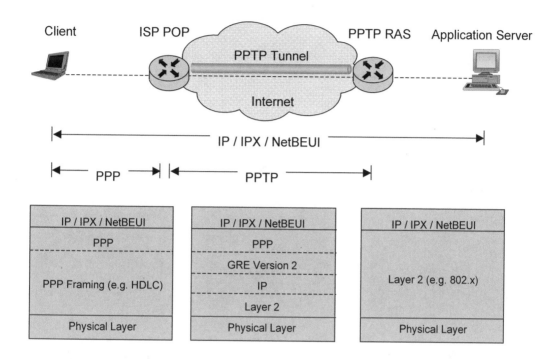

| Client | ISP POP | PPTP Tunnel | PPTP RAS | Application Server |

Internet

IP / IPX / NetBEUI

PPP | PPTP

IP / IPX / NetBEUI	IP / IPX / NetBEUI	IP / IPX / NetBEUI
PPP	PPP	
	GRE Version 2	
PPP Framing (e.g. HDLC)	IP	Layer 2 (e.g. 802.x)
	Layer 2	
Physical Layer	Physical Layer	Physical Layer

The client PC exchanges Layer 3 frames (IP, IPX, NetBEUI, etc.) that are conventionally encapsulated in PPP frames with the dial-up node. The dial-up node encapsulates the received Layer 3 frames in new PPP frames that themselves are encapsulated in GRE frames. These are in turn encapsulated in IP packets and sent to the RAS server. The RAS server decapsulates these packets and routes the contained Layer 3 frames of the client PC to the local subnetwork. Layer 3 frames from the opposite direction are routed in a similar way to the client PC.

Figure 10.11
Compulsory tunnelling with PPTP

Voluntary tunnelling by the client PC creates a complex frame structure on the network segment between the client PC and the dial-up node of the ISP, as shown in Figure 10.12.

Voluntary tunnelling

In this case, a PPTP tunnel is created directly between the client PC and the RAS server. The IP packets being transported for this purpose are again encapsulated in PPP frames between the client PC and the dial-up node of the ISP. The dial-up node extracts the IP packets from the PPP frames and routes them without any further processing to the RAS server. The PPTP tunnel is

Direct PPTP tunnel between PC and RAS server

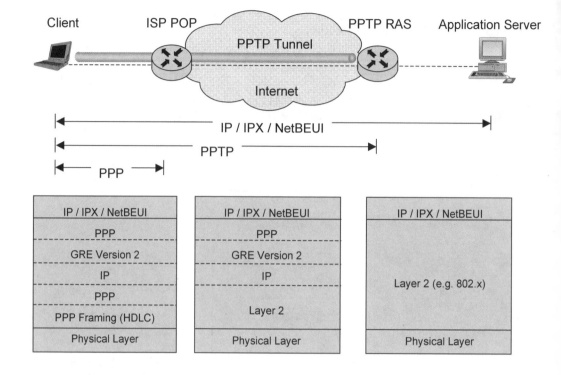

Figure 10.12
*Voluntary tunnelling
with PPTP*

*Frame construction in
client PC*

not visible to the dial-up node because the IP packets of the client PC are not analysed by it.

Figure 10.13 illustrates the construction of the frame structure in the client PC for TCP or UDP-oriented transmission with voluntary tunnelling. First an application process generates user data and sends it with the appropriate system call to the protocol processing interface of the operating system (this normally involves using a *socket interface*). The TCP/IP module then prepares an appropriate TCP or UDP frame and places an IP header in front of it. The PPTP software active in the computer encapsulates this IP packet into a PPP and a GRE frame and arranges for this data to be sent in an IP packet. The IP packet itself is prepared by the TCP/IP module and the appropriate system function is invoked to send the packet. The PPP device driver active in the system encapsulates the IP packet into a PPP frame and sends this frame over the modem connected to the computer to the dial-up node of the ISP. This process is run in 'reverse order' for incoming data with the different protocol headers being processed successively and stripped until the user data can be delivered to the addressed application process.

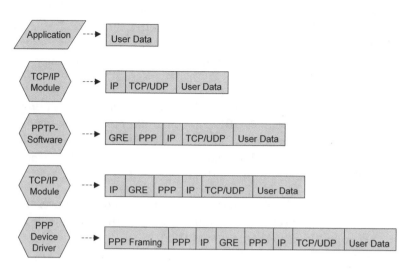

Figure 10.13
*Frame construction
with voluntary PPTP
tunnelling*

10.3.2 Development of PPTP and Alternative Approaches

As a result of Microsoft's support of the protocol in its operating systems, PPTP has enjoyed widespread use during recent years. In fact, Microsoft was heavily involved in the development of the protocol and the corresponding Internet-RFC 2637 [HPV+99]. Microsoft implemented the protocol as part of its *Remote Access Service (RAS)* for the Windows NT operating system and also created appropriate client software for its other versions of Windows.

Widespread use due to Microsoft's support in its operating systems

In addition, Microsoft specified proprietary extensions for the PPP protocol, in particular *Microsoft PPP CHAP Extensions* [ZC98] and *Microsoft Point-to-Point Encryption Protocol* [PZ01].

Proprietary extensions

PPTP does not define any security measures of its own and instead only uses the appropriate protocol mechanisms of the PPP standard or the above extensions proposed by Microsoft. However, not least also because the network-topological usage scenarios were changed from those in PPP, a number of deficiencies were discovered in Microsoft's extensions mentioned above and in PPTP Version 1 as well as in the improved Version 2 [SM98a, SMW99].

No genuine PPTP security measures

At about the same time that PPTP was being developed, Cisco, a company that mainly specialises in network technology, presented a proposal for a competing protocol, the *Layer 2 Forwarding protocol (L2F)* [VLK98]. The IETF working group therefore could not reach a general consensus on adopting PPTP as a standard

Other developments

protocol. As a result, a compromise was reached in which the advantages of both protocols were merged into a single protocol, the *Layer 2 Tunneling Protocol (L2TP)* [TVR+99].

Comparison of PPTP and L2TP

The two protocols (PPTP and L2TP) have the following similarities and differences:

❏ Both protocols use PPP for the initial encapsulation of Layer 3 frames (IP, IPX, NetBEUI, Appletalk, etc.). They also extend the PPP model by allowing Layer 2 end points to reside on other systems than the PPP end points. Both protocols also support voluntary and compulsory tunnelling.

❏ Both protocols offer optional protocol header compression, with L2TP operating with a minimum overhead of four octets compared with six octets for PPTP.

❏ For the communication protocols between the two tunnel end points PPTP requires the transport of GRE-PDUs in IP packets, whereas L2TP supports a range of different technologies, such as IP (with UDP), Frame Relay, X.25 or ATM.

❏ PPTP only supports one tunnel between two tunnel end points (host or gateway system), whereas L2TP allows multiple tunnels simultaneously, therefore enabling separate tunnels to be created for different quality of service requirements of the transmission service.

Virtual Layer 2 connectivity

PPTP as well as L2TP enable *virtual Layer 2 connectivity* between systems that are only able to communicate with each other over a switching network (with Layer 3 involvement). The advantages of virtual Layer 2 connectivity are that the respective systems can use any Layer 3 protocols (IP, IPX, NetBEUI, Appletalk, etc.) to communicate with each other and systems can also be users of the same Layer 3 subnetwork no matter what their actual location is. To a certain extent, this makes it possible to implement *virtual Layer 3 subnetworks*. This idea is considered in more general terms in the following section which presents different definitions and implementations for *virtual private networks (VPN)*.

10.4   Virtual Private Networks

Various definitions

A number of different definitions exist for the term *virtual private networks (VPN)*. Three of the most common definitions are listed [FH98]:

❏ A virtual private network is a private network that is set up within a public network infrastructure (such as the Internet).

❏ A virtual private network is a communications environment in which access to communication services is controlled to permit connections only within a well-defined group of entities. This communications environment is formed through a partitioning of the common underlying communication medium with the communication medium providing its services to the virtual network on a non-exclusive basis.

❏ A virtual private network is a logical computer network with restricted usage that is constructed from the system resources of a relatively public physical network (such as the Internet) with encryption often used and tunnelling links created by the virtual network across the public network.

The last two definitions explicitly incorporate certain security properties such as *controlled access* and *encryption*, whereas the first definition makes no mention of this. Techniques also exist for building virtual private networks that provide minimal security against external attacks. The primary intention with this type of virtual private network lies more in providing uniform logical addressing than incorporating security measures.

A number of different techniques exist for building virtual private networks:

❏ *Based on dedicated connections:* A VPN is established through dedicated Layer 2 connections — for example, using ATM or Frame-Relay connections, 'Multi-Protocol over ATM' (MPOA) or 'Multi-Protocol Label Switching' (MPLS). Depending on the protocol architecture of the underlying connection technology, the security mechanisms for VPNs in this category can be provided efficiently by the protocols used on the links. An example of this is using the *ATM security specification* to secure a VPN [ATM99].

Dedicated connections

❏ *Per route filtering and controlled route propagation:* The basic idea behind this approach is to control the propagation of routing information so that only certain network nodes or subnetworks receive routing information for the other subnetworks of a VPN. The level of security that can be achieved with this approach relies only on the secrecy of certain routing information. It should therefore be classed in the category 'Security by Obscurity' and be avoided because it is

Route filtering

based on attackers not gaining knowledge about the network topology of such a VPN.

Tunnel building

❑ *By building tunnels through the Internet:* Approaches in this category use the protocols GRE, PPP, PPTP and L2TP discussed in this chapter. The *security architecture for the Internet protocol 'IPSec'*, which supplies security services directly in the network layer IP and is discussed in detail in the following chapter, should also be allocated to this category.

In this book virtual private networks are mainly of interest because of the level of security they can provide. Consequently, approaches based on route filtering and controlled route propagation are not given any further attention. Not least because of space, this book mainly focuses on the Internet protocol suite, and approaches based on dedicated connections are therefore not examined. However, interested readers are particularly advised to compare the security protocols handled in this book with the ATM security specification [ATM99]. The connection-orientated communication model available with ATM offers some advantages for the efficient implementation of security measures compared with connectionless Internet technology.

External influence on quality of service, etc.

Irrespective of the degree of security possible in a VPN, other external influences (e.g. quality of service) mentioned in this book cannot be excluded in the case of VPNs that are mainly implemented through tunnel-building in the Internet. The Internet publication *Wired* printed the following fitting comments in its February 1998 issue: '*Sure, it's a lot cheaper than using your own frame relay connections, but it works about as well as sticking cotton in your ears in Times Square and pretending nobody else is around.*'

10.5 Summary

According to the OSI model, the main task of the link layer is to provide reliable data transmission between systems that are directly connected over a medium. It is not responsible for providing security from intentional manipulation. Furthermore, in their functions the communication protocols used in networks today cannot be matched exactly to the OSI layers. As a result, the security protocols of the link layer should be interpreted in the sense that they secure the protocol data units of the communication layer that is transporting the protocol data units of the network layer.

In local area networks, such as Ethernet, Fast Ethernet, token-ring and wireless LANs that are based on IEEE 802 standards, the *IEEE 802.1x* standard currently in development restricts access to the data transmission service of LANs to authenticated and authorised hosts only. For this purpose the standard introduces the basic property of *port-based access control*, which only allows systems attached to a LAN port access to the data transmission service of the LAN if a prior successful authentication and authorisation check has been performed. The standard includes no further specifications for securing data tranmission in LANs from eavesdropping or the modification of transmitted data units. *IEEE 802.1x*

In addition to normal Layer 2 protocol functions, the *Point-to-Point Protocol (PPP)* that was designed for the exchange of Layer 3 protocol data units over serial connections (e.g. dedicated lines, telephone connections) also contains optional protocol mechanisms for authenticating peer entities and protecting transmission from eavesdropping. This involves performing an authentication exchange and using an encryption protocol after the establishment of a Layer 2 connection. The protocol does not however provide for cryptographic security against intentional manipulation or measures against replaying data units. *PPP*

The *Point-to-Point Tunneling Protocol (PPTP)* enables PPP to be used between systems that do not share a serial connection by performing an encapsulation of PPP data units in IP packets. This allows separate systems even from remote network areas to operate in a particular Layer 3 subnetwork. PPTP does not define any new security measures of its own but instead uses the protocol functions that already exist in PPP. The *Layer 2 Tunneling Protocol (L2TP)* that was developed during Internet standardisation combines the advantages of PPTP, which was heavily influenced by Microsoft, and the protocol *Layer 2 Forwarding (L2F)*, which was specified by Cisco. *PPTP*

The approaches mentioned above enable the *virtual private networks (VPN)* to be constructed through tunnel building through the Internet. Other techniques for building virtual private networks include the use of *dedicated Layer 2 connections* and *route filtering*, although the latter is not actually a security measure since its 'security' depends on the secrecy of the routing information. An alternative approach, also based on tunnel building, is the *IPSec security architecture* discussed in the next chapter. *Virtual private networks*

10.6   Supplemental Reading

[ATM99]   ATM FORUM: *ATM Security Specification Version 1.0.*
           February 1999. – AF-SEC- 0100.000

[BV98]   BLUNK, L.; VOLLBRECHT, J.:   *PPP Extensi-
           ble Authentication Protocol (EAP).*   March 1998. –
           RFC 2284, IETF, ftp://ftp.internic.net/rfc/
           rfc2284.txt

[HMNS98]   HALLER, N.; METZ, C.; NESSER, P.; STRAW, M.: *A
           One-Time Password System.* February 1998. – RFC
           2289, IETF, Status: Draft standard, ftp://ftp.
           internic.net/rfc/rfc2289.txt

[HPV+99]   HAMZEH, K.; PALL, G.; VERTHEIN, W.; TAARUD,
           J.; LITTLE, W.; ZORN, G.:   *Point-to-Point Tun-
           neling Protocol.*   July 1999. –   RFC 2637, IETF,
           Status: Informational, ftp://ftp.internic.net/
           rfc/rfc2637.txt

[IEE01]   INSTITUTE OF ELECTRICAL AND ELECTRONICS EN-
           GINEERS (IEEE): *Standards for Local and Metropoli-
           tan Area Networks: Standard for Port-Based Network
           Access Control.* IEEE Draft P802.1X/D11, 2001.

[Kum98]   KUMMERT, H.: *The PPP Triple-DES Encryption Pro-
           tocol (3DESE).* September 1998. – RFC 2420, IETF,
           Status: Proposed standard, ftp://ftp.internic.
           net/rfc/rfc2420.txt

[Mey96]   MEYER, G.:   *The PPP Encryption Control Proto-
           col (ECP).*   June 1996. –   RFC 1968 , IETF, Sta-
           tus: Proposed standard, ftp://ftp.internic.net/
           rfc/rfc1968.txt

[PZ01]   PALL, G.; ZORN, G.: *Microsoft Point-To-Point Encryp-
           tion (MPPE) Protocol.* March 2001. – RFC 3078, IETF,
           Status: Informational, ftp://ftp.internic.net/
           rfc/rfc3078.txt

[Sim94b]   SIMPSON, W.: *The Point-to-Point Protocol (PPP).* July
           1994. – RFC 1661, IETF, Status: Standard, ftp://
           ftp.internic.net/rfc/rfc1661.txt,

[Sim96]   SIMPSON, W.: *PPP Challenge Handshake Authentica-
           tion Protocol (CHAP).* August 1996. – RFC 1994, IETF,

Status: Draft standard, `ftp://ftp.internic.net/rfc/rfc1994.txt`

[SM98a] SCHNEIER, B.; MUDGE: Cryptanalysis of Microsoft's Point-to-Point Tunneling Protocol (PPTP). In: *ACM Conference on Computer and Communications Security*, 1998, pp. 132–141

[SM98b] SKLOWER, K.; MEYER, G.: *The PPP DES Encryption Protocol, Version 2 (DESE-bis)*. September 1998. – RFC 2419, IETF, Status: Proposed standard, `ftp://ftp.internic.net/rfc/rfc2419.txt`

[SMW99] SCHNEIER, B.; MUDGE; WAGNER, D.: Cryptanalysis of Microsoft's PPTP Authentication Extensions (MS-CHAPv2). In: *International Exhibition and Congress on Secure Networking – CQRE [Secure]*, 1999

[TVR+99] TOWNSLEY, W.; VALENCIA, A.; RUBENS, A.; PALL, G.; ZORN, G.; PALTER, B.: *Layer Two Tunneling Protocol (L2TP)*. August 1999. – RFC 2661, IETF, Status: Draft standard, `ftp://ftp.internic.net/rfc/rfc2661.txt`

[VLK98] VALENCIA, A.; LITTLEWOOD, M.; KOLAR, T.: *Cisco Layer Two Forwarding (L2F) Protocol*. May 1998. – RFC 2341, IETF, Status: Historic, `ftp://ftp.internic.net/rfc/rfc2341.txt`

[ZC98] ZORN, G.; COBB, S.: *Microsoft PPP CHAP Extensions*. October 1998. – RFC 2433, IETF, Status: Informational, `ftp://ftp.internic.net/rfc/rfc2433.txt`

10.7 Questions

1. Can you use IEEE 802.1x to protect the integrity and confidentiality of user data exchanged in a local area network?

2. What is the advantage of using the standard IEEE 802.1x to secure local area networks considering that authentication verification is divided between the authenticator and the authentication server? What are the disadvantages from a security perspective?

3. Why is it that with PPP neither DESEv2 nor 3DESE can be used to secure the Password Authentication Protocol?

4. Which packet sizes result in a severer impact of the protocol overhead created by PPTP: small or large IP packets?

5. Why are two PPP packet headers needed for PPTP with voluntary tunnelling?

6. Which of the protocols PPP, PPTP and L2TP protect the data integrity of transmitted user data?

7. What are the advantages of virtual private networks?

11 IPSec Security Architecture

The *IPSec security architecture* comprises a range of protocols and a framework architecture that are used to secure the protocol data units of the *Internet Protocol (IP)*. After a brief review of the basic background of the Internet protocol, this chapter presents an overview of the main components of the IPSec architecture with detailed discussions appearing in the subsequent sections.

11.1 Short Introduction to the Internet Protocol Suite

As a network layer protocol, the Internet protocol has the task of enabling communication between systems that are not directly connected to one another over a shared medium. IP thus offers a connectionless datagram service with no guarantee of packet delivery and is therefore characterised as 'unreliable.' However, this characterisation of IP as being unreliable should not suggest that IP packets are frequently not delivered correctly when the traffic load in the Internet is at a 'normal level.' It only means that the protocol does not provide explicit mechanisms to guarantee correct delivery or any other assurances for quality of service (e.g. end-to-end delay or jitter). The IP service is therefore referred to as *'best effort'*.

Unreliable datagram service

'Best effort' service

Figure 11.1 shows how IP is embedded in the *TCP/IP protocol suite*. The TCP/IP suite organizes the distributed information processing in systems connected over an internetwork (i.e., a 'network of networks') into four protocol layers:

TCP/IP protocol suite

❑ A series of *access protocols* that performs functions up to and including the second OSI layer and which, in practice, are often realised through a LAN protocol, such as Ethernet, token-ring, wireless LAN or, with serial dial-up connections, through PPP.

Access protocols

Figure 11.1
*Distributed
information
processing based on
TCP/IP protocol
suite*

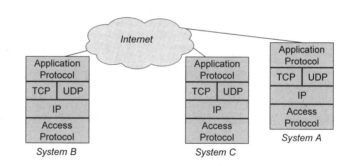

Internet Protocol (IP) ❏ The *Internet Protocol (IP)*, which provides an unreliable connectionless datagram service.

Transmission Control Protocol (TCP) ❏ The *Transmission Control Protocol (TCP)*, which provides a reliable connection-oriented transport service realised over IP.

User Datagram Protocol (UDP) ❏ The *User Datagram Protocol (UDP)*, which offers an unreliable and connectionless transport service and is therefore basically only an application interface to IP with application process addressing.

Application protocols ❏ A range of *application protocols*, including *Simple Mail Transfer Protocol (SMTP)* for the delivery of electronic mail and *Hypertext Transfer Protocol (HTTP)* for transmission of hypertext structures, just to name two of the currently most important ones.

Figure 11.2 shows the structure of an IP packet. The fields contained in the protocol header have the following significance:

Version ❏ *Version (Ver.):* This four-bit field gives the protocol version. Most systems are still using Version 4 of the protocol although specifications for its successor, Version 6, have been available for some time (version number 5 was used for an interim version with an incomplete specification).

Internet header length ❏ *Internet header length (IHL):* This four-bit field gives the length of the IP header in 32-bit words.

Type of service ❏ *Type of service (TOS):* The original purpose of this eight-bit field was to express specific quality of service requirements for packets in the form of prioritisation. It is currently under review by the IETF.

Ver.	IHL	TOS	Length		
IP Identification			Flags	Fragment Offset	
TTL		Protocol	IP Checksum		
Source Address					
Destination Address					
IP Options (if any)					
TCP / UDP / ... Payload					

Figure 11.2
Format of an IP packet

❑ *Length:* The overall length of a packet including packet header and user data is indicated in octets in this 16-bit field. Like all other protocol fields in the TCP/IP protocol suite, this field is coded in 'big endian' representation.

Length

❑ *IP identification:* This field 'uniquely' identifies IP packets and is very important for the segmentation and reassembling of fragmented IP packets.

IP identification

❑ *Flags:* The flag field contains three bits to indicate whether an IP packet can be fragmented or whether an IP packet is already a fragment of a larger fragment and one bit that is reserved for future applications.

Flags

❑ *Fragment offset:* This 13-bit field contains the position of a fragment within a fragmented IP packet.

Fragment offset

❑ *Time to Live (TTL):* At every IP-processing network node this 8-bit field is decremented by one. When the TTL field reaches 0, the packet is discarded at the next network node as long as this node is not the destination node of the packet, thereby preventing packet looping.

Time to Live

❑ *Protocol:* This eight-bit field shows the (transport) protocol of the payload and is therefore used by hosts to identify the correct protocol entity (TCP, UDP, etc.) for the user data.

Protocol

❑ *IP checksum:* The 16-bit checksum contained in this field is used to detect errors in the protocol header. As it is not a cryptographic checksum, it can easily be forged.

IP checksum

❑ *Source address:* This field contains the 32-bit long source address of the sending system.

Source address

❑ *Destination address:* This is also a 32-bit long field and it

Destination address

contains the destination address of the intended receiver of an IP packet.

IP options
❏ *IP options:* An IP packet header can optionally carry variable-length information. As options are not important to IPSec, they will not be discussed further.

Security problems with IP protocol
The Internet protocol has some security deficiencies because security aspects were not included in the requirements for the protocol during its development. An entity receiving an IP packet therefore has no assurances regarding the following security properties:

Data origin authentication and data integrity
❏ *Data origin authentication and data integrity:* Was the packet actually sent by the entity indicated as the sender in the protocol header, and does the packet contain the exact content placed in it by the sender, or was the content of the packet modified while in transit to the receiver? Is the destination address given in the packet header actually the address the sender of the packet specified as the original receiving entity?

Packet replaying
❏ *Replay protection:* Is this packet the actual packet sent or an intentional replay of the packet by a malicious hacker?

Confidentiality
❏ *Confidentiality:* Was the packet spied on by an unauthorised third party while in transit from sender to receiver?

11.2   Overview of IPSec Architecture

IPSec security objectives
The objective of the IPSec security architecture is to eliminate the security problems listed above. Its security objectives are therefore as follows:

❏ *Data origin authentication and data integrity:* It should not be possible to send an IP packet with a forged source address or to modify the destination address of an IP packet without detection by the receiver. It also should not be possible for the content of a packet to be modified without detection by the receiver.

❏ *Replay protection:* It should not be possible for a packet to be replayed at a later time without detection by the receiver.

❏ *Confidentiality:* It should not be possible to read the content of transmitted IP packets. Furthermore, IPSec should provide limited protection from traffic flow analyses.

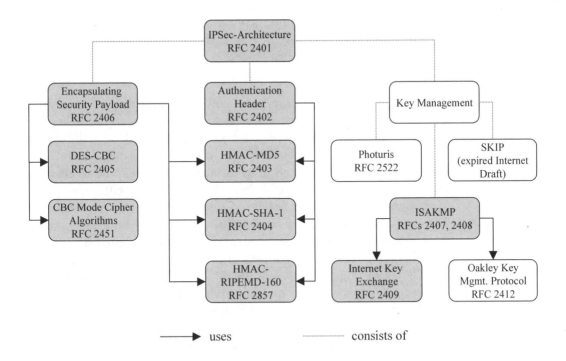

Figure 11.3
Overview of IPSec Standards

Another important objective of the IPSec architecture is to give sender, receiver and all gateway nodes the option of deciding which security requirements they want to apply to specific IP traffic flows according to the local security policy. The sending entities (host and gateway systems) secure IP packets according to a local *security policy* and receiving entities discard packets that have inadequate protection because they lack confidence in these packets.

Security policy

An overview of the key components of the IPSec architecture is presented in Figure 11.3. The main concepts of the architecture are defined in RFC 2401 [KA98c]. These include the concept of *Security Association (SA)*, the conceptual *Security Associations Database (SADB)*, the *Security Policy* as well as the conceptual *Security Policy Database (SPD)*. RFC 2401 also gives an overview of the two fundamental security protocols of the architecture, *Authentication Header (AH)* and *Encapsulating Security Payload (ESP)*, both of which are specified in separate RFCs. Both protocols can be operated in one of two possible operation modes, *transport mode* or *tunnel mode*. Other RFCs specify details on the use of specific cryptographic operations with AH and ESP:

Main concepts and components

❑ DES and other block ciphers in cipher block chaining mode are provided for encryption.

❏ Data origin authentication and integrity protection for IP packets is realised using a HMAC construction, with MD5, SHA-1 and RIPEMD-160 provided as cryptographic hash functions.

The two complementary protocols *Internet Security Association and Key Management Protocol (ISAKMP)* and *Internet Key Exchange (IKE)* are used to negotiate the keys needed for the cryptographic operations above and perform entity authentication.

Security association (SA)
In this context a security association (SA) can be interpreted as a type of 'simplex connection' that provides specific security services to the traffic it carries. The term 'connection' is used with quotation marks because the Internet protocol realises a *connectionless* and thus a *stateless* service for all participating entities. However, the respective peer entities require certain state knowledge to implement the security services, e.g. in respect of the security mechanisms and keys used. This state is administered through security associations established, deployed and dissolved between peer entities. As an SA is basically only used for one communication direction, two security associations are always needed for bidirectional communication.

Identification of SAs
The security services negotiated for an SA are supplied by exactly one of the two security protocols – *Authentication Header (AH)* or *Encapsulating Security Payload (ESP)*. A triple, which consists of a *Security Parameter Index (SPI)*, an IP destination address and the identifier of one of the two security protocols (AH or ESP), provides the unique SA identification for each system.

Peer entities of SAs
A security association can basically be established between the following entities:

❏ host ↔ host

❏ host ↔ gateway

❏ gateway ↔ gateway

Administration and specification of SAs and their parameters
Two conceptual databases should exist in each system for the administration and specification of security associations. The *Security Association Database (SADB)* contains the security associations active in a system at any given time, and the *Security Policy Database (SPD)* defines which security services are being applied to which IP packets and how this application should be executed. The SPD therefore identifies which security associations need to be established between which peer entities for which data streams

and specifies the parameters to be negotiated for these associations. These databases are thus called 'conceptual databases' since no 'real' database technology (e.g. relational database with general query language SQL) is required. The security architecture does not define how the databases are implemented nor are any general database concepts normally used in practice.

A security association is basically operated in one of the following two modes:

❑ *Transport mode* can only be used between the end points of a communication relationship, i.e., between two hosts — or, if a data stream is destined directly for a gateway system (e.g. in the case of network management applications) — between a host and a gateway system.

❑ *Tunnel mode* can be used between any systems.

The difference between the two protocol modes is that transport mode only adds a security-specific protocol header (and possibly a trailer) whereas tunnel mode totally encapsulates the protected IP packets (see also Figure 11.4). This encapsulation of IP packets enables gateway systems to protect certain data streams on behalf of other entities. As a result, these systems can provide uniform protection to entire subnetworks.

Protocol modes

Transport mode

Tunnel mode

IP Header	IPSec Header	Protected Data

(a) Transport Mode

IP Header	IPSec Header	IP Header	Protected Data

(b) Tunnel Mode

Figure 11.4
Packet formats for transport and tunnel modes

The security protocol *Authentication Header (AH)* [KA98a] implements the security service of data origin authentication with replay protection for IP packets. As the name indicates, the protection is provided through a protocol header inserted between the IP protocol header and the user data of the IP packet (see also Figure 11.5).

In contrast, the security protocol *Encapsulating Security Payload (ESP)* [KA98b] offers optional data origin authentication with packet replay protection as well as optional confidentiality for

Authentication header (AH)

Encapsulating Security Payload (ESP)

Figure 11.5

Structure of an IP packet with an authentication header (AH)

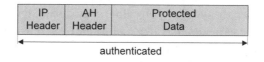

transmitted user data. Selection of at least one of the two optional security services is necessary for an SA. The protocol implementation involves an additional protocol header and a trailer with the user data of the IP packet encapsulated between them (see also Figure 11.6).

Figure 11.6

Structure of an IP packet with Encapsulating Security Payload (ESP)

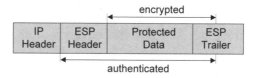

Establishing security associations

The two protocols *Internet Security Association Key Management Protocol (ISAKMP)* and *Internet Key Exchange (IKE)* serve as the basis for the negotiation and establishment of security associations.

Internet Security Association Key Management Protocol (ISAKMP)

The ISAKMP protocol [MSST98] defines a generic framework for entity authentication, key exchange and the negotiation of parameters for security associations. It does not identify a specific authentication protocol, but instead stipulates the 'language' for defining such protocols by specifying fundamental things as packet formats, retransmission timers, message construction requirements, etc. In principle, ISAKMP is defined as a general protocol for the authentication and negotiation of security parameters independently of IPSec. Use of ISAKMP for IPSec is explained in detail in RFC 2407 separately from the protocol definition [Pip98].

Internet Key Exchange (IKE)

The *Internet Key Exchange (IKE)* protocol specified in RFC 2409 [HC98] defines a specific authentication and key exchange protocol that conforms with ISAKMP and may be used for various purposes, although in practice its main usage is the negotiation of security associations for IPSec. This negotiation takes place in two phases: During the first phase an *IKE SA* is established that defines how other security associations for protecting specific data streams should be negotiated between the two peer entities during the second phase.

Replay protection

Both security protocols AH and ESP carry a sequence number

that provides replay protection for IP packets. This sequence number is initialised with the value 0 when an SA is established and incremented by 1 with each packet sent. A new key is needed to replace the session key agreed for the SA before a wraparound of the 32-bit long sequence number occurs. Because this sequence number is included in the computation of the MAC, any modification of it by an attacker will be detected.

The receiver of an IPSec-protected packet always verifies that the sequence number contained in the packet is within a range of acceptable numbers. This range is called a 'sliding window'. The reason an entire window is used is that in the Internet the order of IP packets can change during transmission of IP packets via different routes, even during normal operation. Therefore, later packets may possibly arrive at the receiver sooner than packets that were sent earlier. If receivers were to insist upon a strict sequence of the numbers, they would have to discard IP packets that unintentionally end up in the wrong sequence, which would result in an unnecessary reduction of the data throughput in the Internet. Consequently, a receiver only accepts IP packets if they are not 'too old', i.e., if newer IP packets with significantly higher sequence numbers have not already been accepted. This situation is illustrated in Figures 11.7 and 11.8.

'Sliding window' mechanism

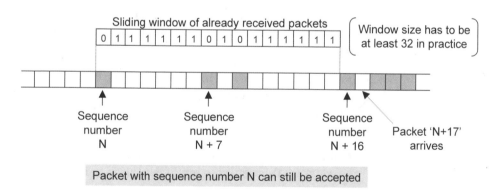

Packet with sequence number N can still be accepted

Figure 11.7 shows an example of a size 16 receiving window. The packets that have not yet been received are indicated in the grey boxes and the vector of the packets that have already been received shows a 0 in these places and a 1 in the other places. Assuming that $N + 15$ is the sequence number of the most recently received packet, the packet with sequence number N is still acceptable. However, if the packet with sequence number $N + 17$ arrives before this packet, the window is advanced by two positions (see

Figure 11.7
Example of a sliding window before updating

Figure 11.8) so that the packet with sequence number N is now placed to the left of the receiving window and therefore discarded as being too old by the time it is received.

Figure 11.8
Example of a sliding window after a window is updated

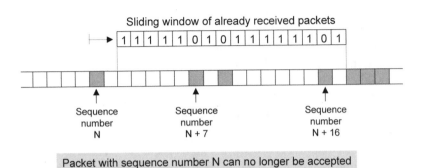

RFC 2401 specifies a minimum window size of 32 with the value 64 recommended as the default. Upon receipt of an IP packet, the receiving system performs the following actions depending on the sequence number:

Actions of the receiver

❏ If the sequence number is to the left of the receiving window, the receiver discards the packet.

❏ If the sequence number is inside the receiving window, the receiver verifies the MAC and accepts the packet if the verification is successful.

❏ If the sequence number is to the right of the receiving window, the receiver verifies the MAC and, if the verification is successful, accepts the packet and advances the window to the right.

IPSec implementation modes

As explained above, IPSec can be implemented in hosts as well as in gateway systems. Two separate implementation modes exist for both cases.

Implementation in a host

The advantages of IPSec implementation in a host are the provision of real end-to-end security services, the availability of specific security services for each individual data stream, and the fact that either transport or tunnel mode can be used.

The two alternatives for integrating IPSec illustrated in Figure 11.9 are also available: *integration of IPSec into the protocol processing functions of the operating system* and *integration as an intermediate layer between the IP layer and the access protocol layer*. The latter mode is mainly used when an operating system

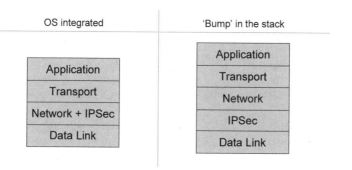

Figure 11.9
Integration alternatives for IPSec in hosts

cannot be modified itself (for instance, when its source code is not available). However, direct integration into an operating system is clearly the method of choice as it avoids a duplication of protocol functionality.

The advantage of IPSec implementation in gateway systems is that IP packet exchange between two subnetworks over the public Internet can be uniformly secured, thereby enabling the construction of *virtual private networks (VPN)*. Furthermore, IPSec implementation and configuration is not necessary in each host in this case and the ability exists for IP traffic flowing between remote users and subnetworks to be uniformly secured and authorised.

Implementation in gateway systems

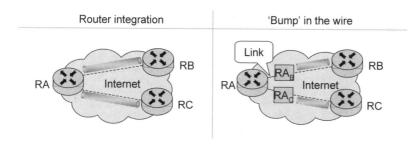

Figure 11.10
Integration alternatives for IPSec in gateway systems

The two alternatives shown in Figure 11.10 are used to implement IPSec in gateway systems: *integration directly into the switching system* and *insertion of external units*. With the first alternative IPSec functionality is directly integrated into the protocol functions of the switching system, whereas with the second alternative, additional IPSec modules are inserted before each protected input or output of a gateway system. Direct integration is preferable for efficiency reasons, so the second alternative is only considered to be a temporary solution for non-IPSec-enabled switching systems.

11.3   Use of Transport and Tunnel Mode

This section takes a detailed look at the differences between transport and tunnel mode and explains specific usage scenarios. First we need to distinguish between the two terms *communication end point* and *cryptographic end point*.

Communication end points and cryptographic end points

Definition 11.1 *The* **communication end points** *of a data stream denote the source and destination system of IP packet exchanges. In contrast, systems that generate and process the AH or ESP protocol headers of IP packet exchanges within the framework of a security association are called* **cryptographic end points**.

Transport and tunnel mode usage

These two terms make it easy to distinguish between the usage areas of transport and tunnel mode:

❑ Transport mode is used when the cryptographic end points are the same as the communication end points (see also Figure 11.11).

❑ Tunnel mode is used when at least one cryptographic end point is not a communication end point (see also Figures 11.12 and 11.13).

Figure 11.11
End-to-end security with transport mode

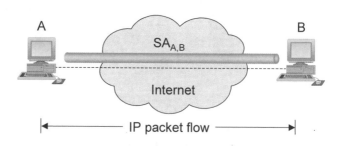

In most cases the communication end points are located in hosts (personal computers, workstations, servers), but not necessarily. For example, if a gateway system is managed by a management station via the Simple Network Management Protocol

(SNMP), then the IP packet exchanges are addressed to or sent by the gateway system.

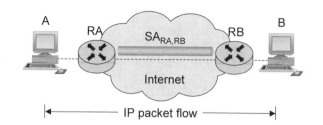

Figure 11.12
Use of tunnel mode in gateway systems

Figure 11.12 illustrates the use of tunnel mode in gateway systems and the structure of packet exchanges between two hosts A and B in the Internet. When packets are transmitted from A to B, the gateway system RA refers to its local security policy to decide whether the packets should be secured based on a security association negotiated with RB. It then generates an appropriate IPSec protocol header (AH or ESP) that is placed in front of the *entire IP packet including the original protocol header*. It also generates a new IP protocol header that is addressed to RB and placed in front of the IPSec protocol header. The original IP packet is therefore transported in a tunnel between RA and RB through the Internet, which also accounts for the term 'tunnel mode'. In practice this sort of configuration is mainly found when two private subnetworks at different locations are networked over the public Internet *(Intranet scenario)*.   *Intranet scenario*

Figure 11.13 shows a different usage scenario for tunnel mode in which the two hosts A and B communicate with one another but this time with the packets being secured between A and RB. Because the cryptographic end point RB is not the same as communication end point B, the packets between A and RB are completely encapsulated. A typical example of this configuration is the *'road warrior scenario'* in which employees of a company have to access the services of the private subnetwork of the company when they are at remote locations (e.g. hotel rooms).   *'Road warrior' scenario*

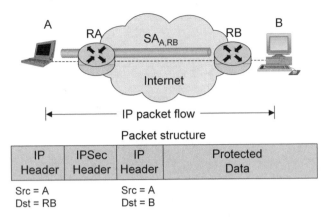

IP Header	IPSec Header	IP Header	Protected Data

Src = A
Dst = RB

Src = A
Dst = B

Nesting of security associations   Security associations can also be nested, as shown in Figure 11.14. A sample application of nesting is when data exchanged between *A* and *RB* is basically authenticated and in addition all data exchanges between *RA* and *RB* are encrypted and, therefore, confidentiality is performed at the subnetwork level. As the illustration also shows, a separate IPSec protocol header and a separate IP protocol header are added to the packets for each security association.

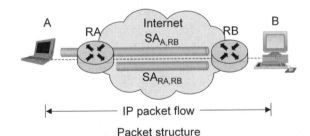

IP Header	IPSec Header	IP Header	IPSec Header	IP Header	Protected Data

Src = RA
Dst = RB

Src = A
Dst = RB

Src = A
Dst = B

It is important when security associations are nested that no overlapping of tunnel segments occurs ('correct bracketing'). Figure 11.15 presents an example of two validly nested associations.

In contrast, the security associations shown in Figure 11.16 are not correctly nested. The packets are transported in a tunnel

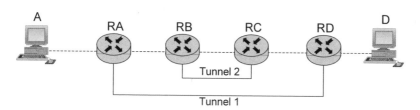

Figure 11.15
Valid nesting of two security associations

Packet structure

IP Header	IPSec Header	IP Header	IPSec Header	IP Header	Protected Data

Src = RB   Src = RA   Src = A
Dst = RC   Dst = RD   Dst = D

between RB and RD in such a way that RC is not able to analyse and strip the IPSec protocol header added by RA for it. Furthermore, after the outer protocol header has been stripped in gateway system RD, the IP packet finds itself at a topologically incorrect position in the network so that it has to be 'routed back' in the direction RC. Depending on the value of its TTL field, it may then be discarded on the way.

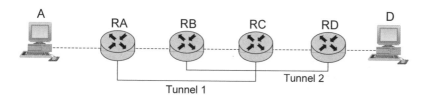

Figure 11.16
Example of two security associations with invalid nesting

Packet structure

IP Header	IPSec Header	IP Header	IPSec Header	IP Header	Protected Data

Src = RB   Src = RA   Src = A
Dst = RD   Dst = RC   Dst = D

11.4 IPSec Protocol Processing

As clarified in the preceding sections, the IPSec architecture allows IP packets to be secured in hosts as well as in gateways. The additional processing procedures required to support IP protocols are listed below. What is evident is that hosts and gateways essentially both provide the same functions so that the only distinction required is between outgoing and incoming packets.

IPSec processing of outgoing packets

Consider a system that has to send a specific IP packet to another system. The following procedures are required for valid IPSec processing:

1. *Determine whether the IP packet has to be secured:* This decision is made on the basis of the local security policy that is stored in the security policy database (SPD). Depending on the action specified in the corresponding SPD entry, the packet is either discarded, sent without any further protection or secured before it is sent.

2. *Determine the security association to be used for the IP packet:* If an SA does not yet exist with the corresponding peer entity, a request for authentication and parameter negotiation is sent to the key management entity.

3. *Read SA parameters:* The parameters of the newly generated SA are read from the SADB.

4. *Perform security measures specified in the SA:* This step results in the generation of an AH or ESP protocol header and a new IP protocol header when tunnel mode is used. Depending on which security services were negotiated for the SA, the related user data is encrypted and/or protected with a MAC. All necessary parameters and keys are stored in the corresponding entry in the SADB.

5. *Start IPSec processing for resulting packet:* Once the IP packet is ready, normal protocol processing continues for all IP packets (see step 1). The reason is that multiple IPSec protocol headers may have to be added in the same system when security associations are nested (look at the direction of B to A in gateway system RB in Figure 11.14).

IPSec processing of incoming packets

The following procedures are performed in a system to process incoming IP packets:

1. *Verify whether the packet contains an IPSec protocol header that the system needs to process:* This verification only considers the most outer IPSec header in each step as IPSec headers always have to be stripped from the outside to the inside.

2. *Process the outer IPSec header:* If the outer IPSec header is stipulated for this particular system, the next step will involve requesting the corresponding SA from the SADB. If the correct SA does not exist in the local system, the packet is dropped. If the correct SA can be found, the packet is processed accordingly. The packet is also discarded if an error occurs (e.g. false MAC).

3. *Decide whether the packet is correctly secured:* The corresponding entry for the packet is searched in the SPD. Depending on the action specified in the security policy, the packet is either discarded or processed. Another check is required to determine whether an additional IPSec header needs to be processed. In addition, state information has to be stored until the packet has been completely processed (payload conveyed to appropriate transport entity, resulting IP packet forwarded to next router or packet discarded). This state information enables a decision on whether the packet was secured as specified in the security policy after all IPSec headers have been stripped.

As the discussion highlights so far, the security policy plays an important role in processing supported by the IPSec protocol. The following *selectors* are the basis for selecting the appropriate entries for IP packets in the SPD:

Selection of appropriate security policy

❏ *IP source address:* This can be the address of an individual system, a network prefix, an address range or a wildcard.

❏ *IP destination address:* This address is specified in the same way as the IP source address. With encapsulated IP packets (tunnel mode), it is always the inner IP packet headers that are evaluated for the source and destination addresses.

❏ *Name:* This can be a 'domain name system' name (DNS), an X.500 name, or another name type as defined in the IPSec-specific definitions for authentication protocols (e.g. RFC 2407 for ISAKMP). Names are only evaluated for the purpose of negotiating security associations.

❏ *Protocol:* This denotes the protocol identification of the transport protocol for the considered packet. ESP-encrypted user data first has to be decrypted before a decision is possible on whether the packet is properly secured. Decryption can naturally only be performed if the system is also the cryptographic end point for this ESP header. Otherwise ESP is assumed to be the transport protocol and it is determined whether the packet can be routed.

❏ *Port of application protocol:* If accessible, the service access point identification of the application protocols can be included in the selection of a security policy.

Content of security policy entries The entry for a packet is selected in the SPD on the basis of the characteristics mentioned. The entries contain the following information:

❏ Details on how a Phase-I-SA (e.g. IKE SA) should be negotiated — for example, with main mode or quick mode — or which protection suites should be used (see Sections 11.7 and 11.8).

❏ Details on which security services should be provided for IP packets. These include:

   ❏ Selectors that themselves identify individual data streams.

   ❏ The executable action for these data streams (discarding, direct routing or security).

   ❏ Security attributes for secured data streams, such as the security protocol (AH or ESP), the protocol mode (transport or tunnel mode), information about the security algorithms and their parameters as well as other parameters such as the lifetime of security associations and the window size for replay protection.

If an SA is already established with a security policy, its identification is referenced in the SPD so that it can be requested efficiently from the SADB.

11.5 The ESP Protocol

ESP security services The ESP protocol is a generic security protocol that provides the following security services for IP packets:

❏ *Confidentiality* through encryption of payload (transport mode) or of complete IP packets (tunnel mode).

❏ *Data origin authentication* and *packet replay protection* achieved through the addition of a MAC and a sequence number.

Both security services are optional and can be combined with one another. However, at least one of the two options must be selected for a security association.

The specification of ESP is divided into two parts: the definition of the base protocol in RFC 2406 [KA98b] and the use of specific cryptographic algorithms with ESP, e.g. encryption with DES-CBC in RFC 2405 [MD98] or with other cryptographic algorithms in CBC in RFC 2451 [PA98] and data origin authentication with HMAC-MD5 in RFC 2403 [MG98a] or HMAC-SHA in RFC 2404 [MG98b].

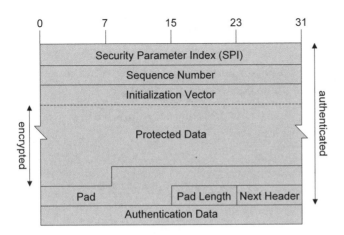

Figure 11.17
Packet format of Encapsulating Security Payload (ESP)

The base protocol definition for ESP in RFC 2406 contains a specification for the protocol header and trailer, fundamental steps for processing the protocol and procedures for transport and tunnel mode. Figure 11.17 shows the packet format for the encapsulating security payload.

The ESP protocol header shown immediately follows an IP protocol header or an IPSec protocol header with the 'next header' field of the preceding protocol header using the value '50' to indicate that it is an ESP protocol header. The protocol fields of ESP have the following significance:

ESP protocol fields

❏ The *Security Parameter Index (SPI)* field identifies which SA should be used for a packet. The value entered in this field is always determined by the receiving side during SA negotiation because it is the receiver that must provide unique identification of the SA based on the SPI.

Security Parameter Index (SPI)

❏ As explained before, the *sequence number* provides replay protection for IP packets.

Sequence number

Initialisation vector (IV)   ❑ If the cryptographic algorithm being used requires an *initialisation vector (IV)*, the IV is transported in plaintext in each IP packet so that each packet can be processed independently of other packets.

Pad   ❑ The *Pad* field ensures that the payload being encrypted is padded to a length that is equivalent to an integer multiple of the block size of the algorithm used and that the two following fields end up in the higher-order 16 bits of a 32-bit word.

Pad length   ❑ The *Pad Length* field indicates the number of octets added.

Next header   ❑ The *next Header* field indicates the protocol type of the contained payload, e.g. TCP or UDP.

Authentication data   ❑ The optional *authentication data* field contains a message authentication code (MAC), if available.

Protocol processing of outgoing packets   Figures 11.18 and 11.19 show the processing procedures for preparing outgoing ESP packets.

Figure 11.18
Preparation of outgoing ESP packets (1/2)

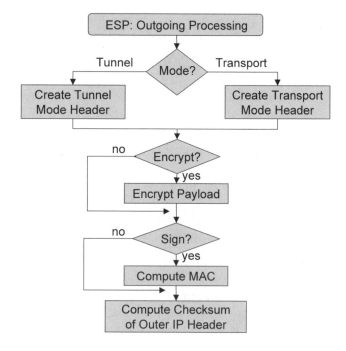

Based on the SA specified by the security policy, a determination is first made about whether to protect the packet in transport or in tunnel mode. Once the appropriate protocol headers have been generated, it has to be determined whether the packet is to be encrypted, and if so, encryption is performed. Verification is also required to decide whether the packet should be authenticated, and if so, a MAC is computed. Lastly, the checksum of the outer IP protocol header is computed.

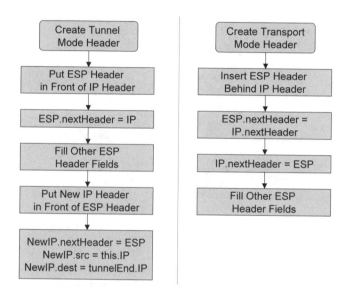

Figure 11.19
Preparation of outgoing ESP packets (2/2)

Figure 11.19 illustrates how protocol headers are prepared for transport and tunnel mode. In tunnel mode an ESP protocol header with the next-header field set to 'IP' is placed before the existing IP protocol. Once the other ESP-specific fields are filled, a new IP protocol header is placed before the ESP header, which has its next-header field set to 'ESP' and its source and destination fields filled with the addresses of the two tunnel end points.

Preparation of protocol headers

In transport mode an ESP header is inserted between the IP header and the payload. The next-header field of the ESP header is set to the value of the corresponding field in the IP header before the next header field of this header is set to the value 'ESP'. Afterwards the remaining ESP fields are filled.

Both Figures 11.20 and 11.21 show the procedures for processing incoming ESP packets.

Protocol processing of incoming packets

Verification is required with incoming packets to ensure that all fragments belonging to a specific packet are already available because processing cannot commence until all fragments have been received. Then the SA being used for the packet is read or the

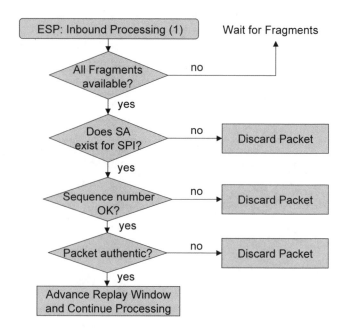

packet is discarded if this association does not exist locally. A further check is applied to ensure that the sequence number is in the window or to the right of the window of acceptable sequence numbers. If not, the packet is discarded. Depending on the result of the MAC verification, the packet is either discarded or the replay window is advanced to the right and the processing of the packet continued.

Figure 11.21 shows how the processing of the packet continues. The payload of the packet is first decrypted and then checked to determine the mode in which the packet was secured. If it was in tunnel mode, then only the two outer protocol headers are stripped. If it was in transport mode, the original value of the next-header field is restored in the IP protocol header before the inserted ESP header is removed and the checksum of the IP protocol header is recomputed. With both modes the packet is checked to determine whether the security measures stipulated in the security policy were used to protect it before it is either discarded or conveyed to the appropriate protocol entity.

*Handling of
fragmented packets*   An IP packet processed and reconstructed in this way can sometimes turn out to be a fragmented packet. This can occur, for instance, if a gateway system applies ESP in tunnel mode to already fragmented packets. A gateway system cannot fully de-

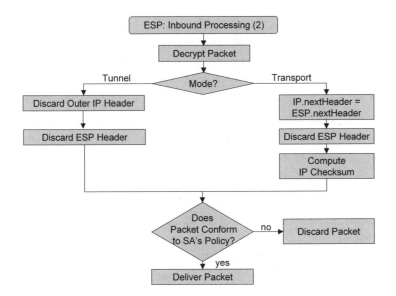

Figure 11.21
Processing of incoming ESP packets (2/2)

cide how to treat a fragmented packet in accordance with the local security policy unless it has actually received all fragments of the packet or at least the fragments that contain the parts of the packet being evaluated (e.g. up to and including the port fields of the transport protocol header). It is possible that only those packets that are being sent to a specific port may be exchanged within an SA. The port information that needs to be evaluated for this purpose is only available in the first fragment of the IP packet.

The following alternatives are available for routing received packets in the context of ESP processing:

Routing alternatives

❏ If another IPSec protocol header is detected that requires attention by this system then IPSec processing is continued.

❏ If a packet was secured in tunnel mode and the system performing the processing is not the communication end point of the packet, then the packet is routed according to local routing information.

❏ If the received packet is destined for the system itself, it is forwarded to the appropriate transport entity (e.g. TCP or UDP).

If within the framework of an SA an ESP provides confidentiality as well as data origin authentication, it is possible for both security services to use different keys if this has been agreed during the negotiation of the SA.

Use of cryptographic algorithms

As explained earlier, the ESP specification is divided into a definition of the base protocol and a description of the use of specific cryptographic algorithms with ESP. The following RFCs were approved:

❏ *Confidentiality:* The use of DES in CBC mode specified in RFC 2405 [MD98] is no longer being recommended because of the short key length (however, this deficiency was already general knowledge when the RFC was approved in 1998). RFC 2451 [PA98] defines format and processing procedures for the use of various block ciphers in CBC mode with ESP. In principle, each block cipher with a 64-bit block size can be used according to this RFC. The algorithms explicitly mentioned are Blowfish, CAST-128, 3DES, IDEA and RC5. The initialisation vector is transmitted in plaintext in each packet to avoid synchronisation problems (see also Section 2.5). The first initialisation vector should be selected randomly, whereas the other vectors can be formed from the last computed ciphertext.

❏ *Data origin authentication:* The same algorithms are used to compute MACs for ESP as for AH (see also the section below).

11.6   The AH Protocol

AH security services

The *authentication header (AH)* specified in RFC 2402 [KA98a] is a generic security protocol that provides the following security services for IP packets:

❏ *Data origin authentication:* A MAC is computed and added to each protected packet.

❏ *Replay protection for IP packets:* The acceptance of a sequence number in the AH protocol header and the inclusion of this number in the MAC computation enable a receiver to check the freshness of the received IP packets.

As with ESP, the AH specification is divided into two parts: a definition of the base protocol and use of cryptographic algorithms with AH. The protocol specification [KA98a] defines the format of the protocol header, the fundamental processing procedures and operation in transport and in tunnel mode.

If both ESP and AH are to be applied to an IP packet by the same system, then ESP should always be applied first. This makes AH the outer of the two protocol headers and also enables it to protect the ESP header from modification. As an SA basically can only use one of the two protocols AH or ESP, two separate security associations are required per direction.

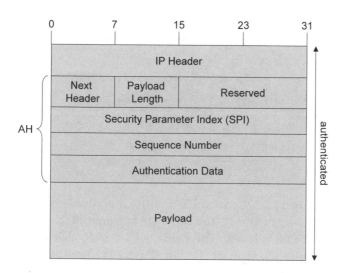

Figure 11.22
Packet format of authentication header (AH)

Figure 11.22 shows the structure of an IP packet secured with AH and the AH protocol fields:

AH protocol fields

❑ The *next-header* field contains the protocol identification of the payload transported in the packet (e.g.TCP, UDP).

Next-header

❑ The length of the AH payload is contained in the *payload length* field. This easily confused labelling refers to the two integrity-securing fields *sequence number* and *authentication data* that, depending on the cryptographic algorithm used and the negotiated parameters, can in principle be of various lengths.

Payload length

❑ The protocol field labelled *reserved* is reserved for future uses and set to 0.

Reserved

❑ The *security parameter index (SPI)* together with the IP destination address uniquely identifies the SA used for this packet.

Security parameter index (SPI)

❑ As explained above, the *sequence number* is used to detect intentional packet replay.

Sequence number

Authentication data

❏ The *authentication data* field contains the MAC that is computed over the entire IP packet including parts of the IP protocol header. In contrast to ESP, AH is also able to protect the immutable parts of the outer IP protocol header from modification.

Figure 11.23
Variable and immutable fields of an IP packet header

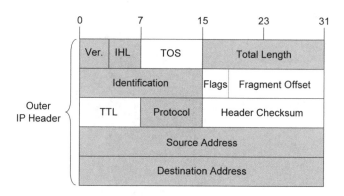

In Figure 11.23 the variable and the immutable fields of an IP protocol header are distinguished from one another through different background colours. The fields with a white background can change during the transport of an IP packet from source to destination (type of service, flags, fragment offset) or in principle are subject to change in each gateway (time to live, header checksum). These fields naturally cannot be included in the MAC as the IP packets would otherwise inevitably be discarded by the receiver. The value 0 is therefore assumed for variable fields in the MAC computation.

Protocol processing of outgoing packets

Figures 11.24 and 11.25 show the procedures for processing outgoing IP packets secured with AH.

As with ESP protocol processing, the packet is checked to determine whether it should be secured in transport or in tunnel mode. After the appropriate protocol headers are created, the message authentication code is prepared and then the checksum of the outer IP protocol header is recomputed.

The protocol headers shown in Figure 11.25 are created using the same scheme as with ESP. First an AH protocol header is placed before the existing IP header and the next-header field of the AH header is set to 'IP'. After the other AH fields are filled, a new IP header is created and placed in front of the AH header. The next-header field of the new IP header is set to the value 'AH' and its source and destination address fields are filled with the addresses of both cryptographic end points.

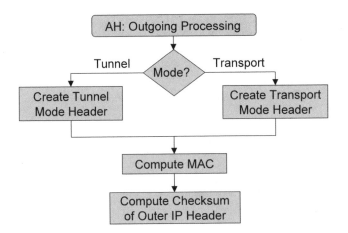

Figure 11.24
Preparation of outgoing AH packets (1/2)

The processing of incoming packets secured with AH is illustrated in Figures 11.26 and 11.27.

Protocol processing of incoming AH packets

When processing incoming packets, the first thing to determine is whether all fragments for the AH protocol data unit being processed are available since processing cannot continue until all fragments have been received. When all fragments are available, clarification is needed to establish whether the SA referenced in the packet is available locally, and if it is not, the packet is dropped.

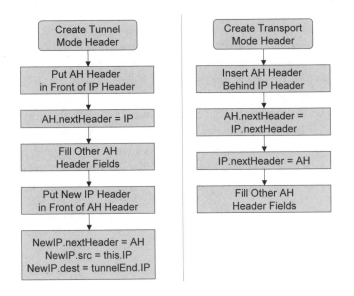

Figure 11.25
Preparation of outgoing AH packets (2/2)

Figure 11.26
*Processing of
incoming AH packets
(1/2)*

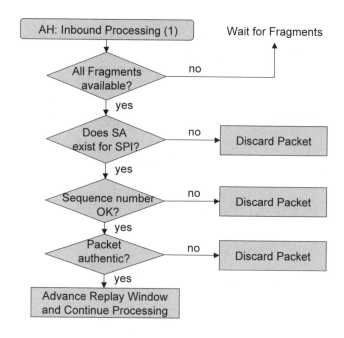

Figure 11.26 *Processing of incoming AH packets (1/2)*

Afterwards the currentness of the sequence number contained in the packet is checked and the packet dropped if this number is to the left of the current receiving window. The message authentication code is then computed and compared with the contained value. If the two values differ, the packet is dropped. The receiving window may then be updated and the processing is continued.

If the security association concerned is operating in tunnel mode, the outer IP header and the AH header are stripped. In the case of a transport mode SA, the next-header field of the IP header first has to be restored before the AH header can be removed. The checksum of the IP header is then recomputed. Finally, the packet is checked to ensure that it complies with the requirements of the local security policy. If it does, it is routed to the appropriate protocol entity; otherwise it is discarded.

Cryptographic algorithms   Three RFSs have been approved for computing the message authentication codes that are inserted into AH and ESP packets:

❏ RFC 2403 defines the construction HMAC-MD5-96 with a key length of 128 bits [MG98a].

❏ RFC 2404 defines the construction HMAC-SHA-1-96 with a key length of 160 bits [MG98b].

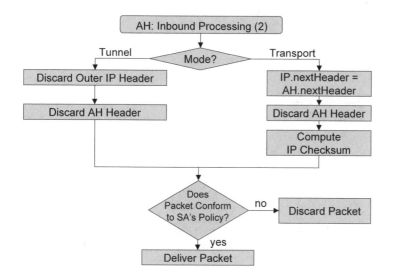

Figure 11.27
Processing of incoming AH packets (2/2)

❏ RFC 2857 defines the construction HMAC-RIPEMD-160-96 with a key length of 160 bits [KP00].

All the above message authentication codes use the HMAC construction defined in RFC 2104 [KBC97]. The suffix '-96' in the algorithm means that the output of the hash construction is truncated to the higher-order 96 bits. This value fulfills the security requirements stated in RFC 2104 and allows a reduction in the protocol overhead. It is therefore defined as the default authentication length of the AH [KA98a].

11.7 The ISAKMP Protocol

Before data packet exchanges between two systems can be protected with IPSec, two diametral security associations are set up between the respective cryptographic end points. These IPSec security associations can basically use one of two options, either *manual* or *dynamic establishment*. Manual establishment involves the manual use of system management methods, which are normally implemented differently from system to system. As this method is not only time consuming but also error prone, it should only be used in very manageable configurations — for example, between two dedicated gateways at different locations. It is also suitable in cases where a dynamic method is not available to a

Manual establishment of SAs is time consuming and error prone

specific system, although this situation should be occurring less and less.

IPSec also defines a standardised method for the dynamic establishment of security associations. This is handled through the following two protocol specifications:

ISAKMP

❏ The *Internet Security Association and Key Management Protocol (ISAKMP)* [MSST98] defines generic protocol formats and procedures for the negotiation of security parameters and for entity authentication. The actual application of this protocol for the negotiation of parameters for IPSec security associations is presented in detail in the *IPSec Domain of Interpretation (IPSec DOI)* [Pip98].

IKE

❏ *Internet Key Exchange (IKE)* [HC98] defines the standard authentication and key exchange protocol for IPSec.

ISAKMP is discussed in detail below. Internet key exchange is covered in Section 11.8.

The ISAKMP specification defines two basic categories of exchanges: *Phase 1 exchanges*, which are used to negotiate *master SAs*, and *Phase 2 exchanges*, which use master SAs to establish other security associations.

Figure 11.28 shows the basic structure of an ISAKMP message. It contains the following message elements:

Initiator and responder cookie

❏ *Initiator and responder cookie* uniquely identify an ISAKMP exchange or an ISAKMP-SA and also provide limited protection against 'denial of service' attacks (explanation follows below).

Next payload

❏ *Next payload* specifies which ISAKMP payload type follows the protocol header.

Major and minor version

❏ *Major and minor version* identify the protocol version of the current message.

Exchange type

❏ *Exchange type* indicates the type of ISAKMP exchange that is conducted with the current message. There are five predefined generic exchange types; other types can be negotiated in DOI specifications.

Flags

❏ *Flags* contain bits to indicate specific characteristics: The *encrypt* bit indicates whether the payload following the message header is encrypted; the *commit* bit indicates a key change; and the *'authenticate only'* bit indicates that the payload of the message is authenticated but not encrypted.

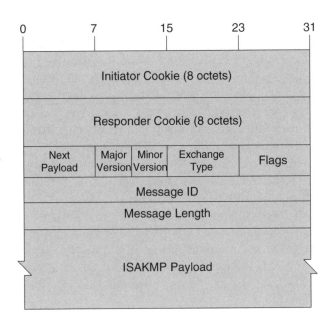

Figure 11.28
Frame format of ISAKMP data units

❏ *Message ID* identifies messages that belong to different protocol runs and therefore allows a simultaneous negotiation of multiple security associations.

Message ID

❏ *Message length* indicates the total length of the current message including the ISAKMP protocol header and all payloads.

Message length

❏ *ISAKMP payload* contains the message payload. An ISAKMP message can actually contain multiple ' chained' payloads. The payload type of the following message is always indicated in the next payload field of the preceding payload or in the ISAKMP protocol header.

Payload

In addition to identifying a protocol run or an ISAKMP-SA, the two message elements *initiator cookie* and *responder cookie* provide limited protection against *'denial of service'* attacks that are aimed at reducing system availability. The motivation for this protection mechanism is as follows: Depending on the authentication scheme used, authentication and key management can involve some very computationally intensive operations, e.g. exponentiation with large integers for the Diffie–Hellman protocol. By deliberately provoking such computations, attackers can cause such an increase in system load that the system is not available to

Protection from 'denial of service' attacks

deal with other tasks or with authentication requests from authorised systems. Attackers then try to keep their own system load
as low as possible, e.g. by only participating in a protocol run until they have triggered a computationally intensive operation in
the attacked system. These simulated authentication exchanges
are consequently not successful. To a limited degree a system can
protect itself from such attacks by not responding to a specific system's requests after a certain number of aborted attempts have
been made and only sending a local error message.

Protection against requests with forged source addresses

However, this is not an effective measure if an attacker produces a request using a forged source address, which then triggers
a computationally intensive operation. The two 'cookies' contained
in ISAKMP protocol headers can be used to reduce these particular risks. Each of the two systems participating in a protocol run
sends the other system information that only the latter can generate and quickly verify, and only performs computationally intensive operations if the peer entity includes this information in its
protocol messages:

❏ The initiating ISAKMP entity creates an initiator-cookie:

$$CKY-I := H\left(Secret_{Initiator}, Address_{Responder}, t_{Initiator}\right)$$

❏ The responder creates its own cookie:

$$CKY-R := H\left(Secret_{Responder}, Address_{Initiator}, t_{Responder}\right)$$

Each of the two entities inserts both cookies into its protocol messages and verifies its own cookie before performing a computationally intensive operation. This protection mechanism is effective at fending off attacks with forged source addresses described
above. An attacker will be unable to intercept the message while it
is being transported between the attacked system and the forged
source address if he or she does not have a valid cookie for this address. The ISAKMP protocol specification defines an exact method
for creating cookies. However, it is important that each entity
has the information necessary to verify its own cookies (e.g. time
stamp).

ISAKMP protocol header

All ISAKMP payloads begin with the protocol header shown in Figure 11.29:

❏ The *next payload* field indicates the type of the next payload.

❏ The *reserved* field is reserved for future applications.

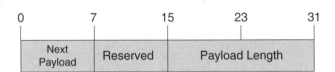

Figure 11.29
*Protocol header for
an ISAKMP payload*

❑ The *payload length* field contains the total length of the payload including the protocol header.

The ISAKMP specification defines a range of different payloads (see [MSST98] for a complete list):

ISAKMP payloads

❑ *Generic payloads* are of general usability, such as the payloads *hash, signature, nonce, vendor ID* and *key exchange*.

❑ *Specific payloads* satisfy certain functions within an authentication exchange and include the payloads *SA, certificate, certificate request* and *identification*.

❑ *Dependent and encapsulated payloads* are only used in conjunction with certain other payloads and provide further detail about the corresponding payload. Examples include:

  ❑ *Proposal payload*, which specifies a proposal for SA negotiation.

  ❑ *Transform payload*, which is part of a proposal payload and describes a specific security transformation.

❑ The *generic attribute payload* is not a self-contained ISAKMP payload and only occurs within other payloads.

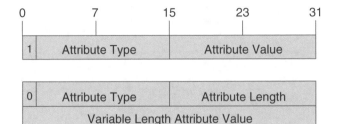

Figure 11.30
*Short and long format
for ISAKMP
attributes*

All attribute payloads share the same structure as shown in Figure 11.30. Short attributes, where the length does not exceed 16 bits,

*Structure of attribute
payloads*

are displayed with a '1' set in the highest-order bit. This is followed by a 15-bit long specification of the *attribute type* and then a 16-bit long *attribute value*. Longer attributes are indicated with an '0'. With long attributes, the type specification is followed by an *attribute length* field to indicate the length of the following *variable length attribute value*.

Figure 11.31
Structure of SA payload

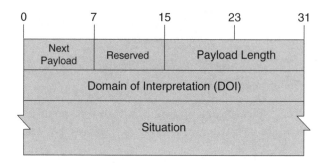

Figure 11.31 shows the structure of an SA payload, which initiates the specification of the desired SA parameters. The *Domain of Interpretation (DOI)* field, which follows the standard payload header, always carries the value 1 defined in RFC 2407 for the negotiation of IPSec-SAs. The *situation* field identifies the specific situation under which a current request is taking place, such as 'normal' or for 'emergency calls'. So far situations have not played an important role in IPSec usage. A series of proposal payloads always follow an SA payload.

Figure 11.32
Structure of the proposal payload

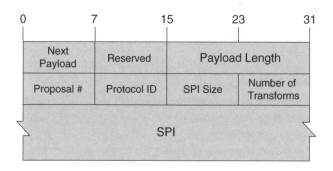

The structure of a proposal payload is illustrated in Figure 11.32. The *proposal #* field is used to communicate the security policy *Linking multiple* and negotiate separate proposals. If two or more proposal pay-*proposals* loads carry the same number, it displays a logical 'AND'. If the proposal numbers have different values, a logical 'OR' is triggered with descending priority. The *protocol ID* specifies which security protocol should be used for the current negotiation (with IPSec: AH or ESP). The *SPI size* field indicates the length of the Security Parameter Index (SPI) and *number of transforms* shows how many transform payloads belong to the current proposal. The corresponding transform payloads directly follow the proposal payload.

The format of a transform payload is presented in Figure 11.33. This payload specifies which security mechanisms – called transformations – should be used within the context of a security association (or a proposal for it). Each transformation listed within a proposal carries a unique *transform #* and a *transform ID*, e.g. for DES, 3DES, MD5 or SHA, to identify its type. Transform IDs are defined in the DOI specifications (for IPSec RFC 2407 [Pip98]). Specific attributes of transformations can be defined in the *SA attributes* field.

Tranform payloads specify which security mechanism should be used

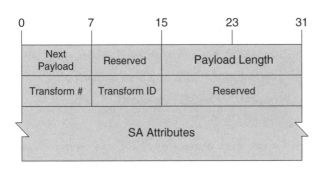

Figure 11.33
Structure of the transform payload

The message formats introduced above will be used to explain how different options for IPSec-SAs are negotiated. The proposal payload allows an initiating ISAKMP entity to notify its peer entity of the security measures it supports or that are required for the currently negotiated SA and to do so in the sequence of its preference. The same proposal number can be used to combine multiple security mechanisms into a closed *protection suite*. However, these proposal payloads must also be considered as a unit within the ISAKMP message and cannot be separated by a proposal payload carrying a different payload number.

Negotiation of protection suites

The following two examples help to illustrate the basic idea involved:

Examples

1. Assume that an initiator wants to negotiate a combined protection suite using ESP and AH, selecting either 3DES or DES for ESP encryption and using the SHA algorithm for

integrity protection. The initiator inserts the SA payload followed by the proposal and transform payloads into the ISAKMP message:

❏ [Proposal 1, ESP, (Transform 1, 3DES, ...),
   (Transform 2, DES)]

❏ [Proposal 1, AH, (Transform 1, SHA)]

If this proposal is accepted by the responder, then two security associations result in each direction: one for ESP and one for AH. The IP packets exchanged within the framework of these associations are protected with 3DES and SHA or with DES and SHA.

2. Two different protection suites should be proposed for the second example. The first suite requires the use of AH with MD5 and ESP with 3DES, whereas the second suite only requires ESP with either 3DES or DES. The initiator sends the following proposal and transform payloads:

❏ [Proposal 1, AH, (Transform 1, MD5, ...)]

❏ [Proposal 1, ESP, (Transform 1, 3DES, ...)]

❏ [Proposal 2, ESP, (Transform1, 3DES, ...),
   (Transform 2, DES, ...)]

Again for the first protection suite two security associations are required per direction, whereas with the second one only one SA has to be established per direction.

Note that the specification of transformations does not allow two different transformations (e.g. 3DES and DES) to be defined simultaneously for an instance of the protocol specification (e.g. ESP). The transform payloads within a proposal payload therefore always carry different numbers.

The responder selects a proposal   The responder selects a proposal from those offered and responds with an ISAKMP message that (among other things) comprises an SA payload and the proposal and transform payloads of the selected proposal. The proposal payload can only contain a transform payload which originates from the list of transformations previously offered to this proposal. To simplify the protocol processing for the initiator, the responder should also select the same numbers for the proposal and transform payloads as appeared in the original message of the initiator. Upon receipt of this message, the initiator verifies that the returned proposal and transform payloads match the combination it used in its own message before it continues with the protocol processing.

Four different session keys are negotiated with ISAKMP for a Phase 1 exchange (see Section 11.8):

Negotiated session keys

- ❏ *SKEYID* is a string only known to the two entities actively involved in the authentication and is used as the *master key* to derive additional keying material. The computation of this key depends on the authentication method.

- ❏ *SKEYID_e* is the encryption key with which ISAKMP entities encrypt their payloads to protect the confidentiality of their messages (the 'e' denotes encrypt).

- ❏ *SKEYID_a* is used by ISAKMP entities to authenticate protocol message exchanges (the 'a' denotes authenticate).

- ❏ *SKEYID_d* is the keying material used to derive keys for non-ISAKMP security associations (the 'd' denotes derive).

11.8 Internet Key Exchange

Based on the message formats and procedures provided by ISAKMP, *Internet Key Exchange (IKE)* specifies the standard protocol for negotiating IPSec security associations. IKE defines the following exchange types, which can be divided into three categories:

- ❏ *Phase 1 exchanges* used to negotiate IKE-SAs:

  Phase 1 exchanges

  - ❏ *Main mode exchange* negotiates an SA through the exchange of six messages.

  - ❏ *Aggressive mode exchange* only requires three messages to do the same thing but cannot guarantee certain security properties of main mode exchange (see below).

- ❏ *Phase 2 exchanges* used to negotiate IPSec security associations:

  Phase 2 exchanges

  - ❏ *Quick mode exchange* exchanges three messages to negotiate additional IPSec associations based on an existing IKE-SA.

- ❏ Other exchanges for various purposes:

  Other exchanges

❏ *Informational exchange* is used to communicate status and error messages between two IKE entities.

❏ *New group exchange* is used to negotiate a new Diffie–Hellman group.

Computation of session keys

IKE provides detailed specifications for computing the four session keys described in the ISAKMP specification and listed in the previous section. *SKEYID* computation, on the other hand, is dependent on the actual authentication method and is therefore explained below. The following definitions are valid for the other three session keys:

$$SKEYID\_d := H(SKEYID, g^{x \cdot y}, CKY{-}I, CKY{-}R, 0)$$
$$SKEYID\_a := H(SKEYID, SKEYID\_d, g^{x \cdot y},$$
$$CKY{-}I, CKY{-}R, 1)$$
$$SKEYID\_e := H(SKEYID, SKEYID\_a, g^{x \cdot y},$$
$$CKY{-}I, CKY{-}R, 2)$$

The $g^{x \cdot y}$ denotes the negotiated shared Diffie–Hellman secret and $CKY{-}I$ and $CKY{-}R$ denote the initiator or responder cookie.

'Expanding' keys

If the key length of a cryptographic algorithm requires more bits than are computed by the hash function, the keys mentioned above are 'expanded' according to the following rule:

$$K := (K_1, K_2, ...) \text{ with } K_i := H(SKEYID, K_{i-1}) \text{ and } K_0 := 0$$

11.8.1   Negotiation of an ISAKMP-SA

Authentication based on hash values

IKE Phase 1 exchanges are authenticated using two hash values computed by the *initiator* and the *responder* if the appropriate message elements exist:

$$Hash{-}I := H(SKEYID, g^x, g^y, CKY{-}I, CKY{-}R,$$
$$SA{-}offer, ID{-}I)$$
$$Hash{-}R := H(SKEYID, g^y, g^x, CKY{-}R, CKY{-}I,$$
$$SA{-}offer, ID{-}R)$$

In the computation above, g^x and g^y denote the two exchanged public Diffie–Hellman values, $ID{-}I$ and $ID{-}R$ the identity of the initiator and the responder, and $SA{-}offer$ the SA, proposal and transform payloads exchanged during the SA parameter negotiation. IKE supports four methods of authentication for peer entities and message exchanges:

❏ *Pre-shared key:* With this method the initiator and the responder know a shared secret $K_{Initiator,Responder}$ which, together with the random numbers $r_{Initiator}$ and $r_{Responder}$ selected by both entities, is included in $SKEYID$ computation:

$$SKYEID := H(K_{Initiator,Responder}, r_{Initiator}, r_{Responder})$$

❏ *Public key encryption:* Public key encryption is supported in two methods where the $SKEYID$ master key is always computed in the same way:

$$SKEYID := H(H(r_{Initiator}, r_{Responder}), CKY-I, CKY-R)$$

❏ *Digital signature:* With this method the two hash values Hash-I and Hash-R also have to be signed by the initiator or the responder because $SKEYID$ itself provides no authentication. The value $SKEYID$ is computed in this case as follows:

$$SKEYID := H(r_{Initiator}, r_{Responder}, g^{x \cdot y})$$

The different versions of Phase 1 exchanges are briefly presented below and distinguished according to the differences between them.

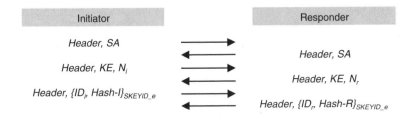

Figure 11.34
Process for main mode exchange with pre-shared key

Figure 11.34 illustrates the process for the main mode exchange with authentication using a pre-shared key. For space reasons the IKE notation N_i and N_r is used for the random numbers ($r_{Initiator}$, $r_{Responder}$) selected by the initiator and the responder. In addition, $Header$ denotes the ISAKMP protocol header, SA the exchanged SA, proposal and transform payloads, ID_i and ID_r the identity of the initiator and the responder and KE (denoting key exchange) the exchanged public Diffie–Hellman values.

Authentication with pre-shared key

The messages exchanged are authenticated on the basis of the hash values in the last two messages. Because the value SKEYID is included in the computation of both Hash-I and Hash-R, after checking these hash values both peer entities can be assured that the respective other entity also knows the SKEYID value. However, because knowledge of the shared secret $K_{Initiator,Responder}$ is required for the computation of SKEYID, each entity is able to conclude that its peer entity also knows this secret and therefore is authentic. As all key message elements of the prior message exchange — including the two current random numbers N_i and N_r — are included in the computation of both hash values, both sides have the assurance that none of the messages exchanged was forged, modified or replayed by an attacker.

Identity confidentiality

The encryption of the ISAKMP payloads in the last two messages also protects the two identities ID_i and ID_r from possible eavesdropping. This measure is particularly useful if the identities cannot be derived from the source and destination addresses of the IP packets that are transporting the ISAKMP messages (e.g. with delegated SA negotiation through security gateways).

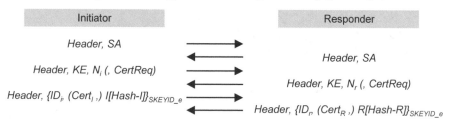

Initiator		Responder

Header, SA   ⟶

⟵   Header, SA

Header, KE, N_i (, CertReq)   ⟶

⟵   Header, KE, N_r (, CertReq)

Header, {ID_i, (Cert$_I$,) I[Hash-I]}$_{SKEYID\_e}$ ⟶

⟵   Header, {ID_r, (Cert$_R$,) R[Hash-R]}$_{SKEYID\_e}$

(m)   denotes that *m* is optional
I[m]   denotes that *I* signs data *m*

Figure 11.35
Process of main mode exchange with signatures

Figure 11.35 illustrates the message flow in main mode exchange with authentication using digital signatures. The contained certificate requests ($CertReq$) and responses ($Cert_I$, $Cert_R$) are in parentheses because they are optional. The notation $I[m]$ and $R[m]$ identifies that message element m is signed by the initiator or the responder. This type of exchange requires explicit digital signatures for the actual authentication procedure because no knowledge of specific keys is needed to compute the two hash values. The two peer entities therefore cannot rely on the hash values to conclude whether an attacker has computed them. The encryption of the ISAKMP payloads in the last two messages protects the confidentiality of the two identities.

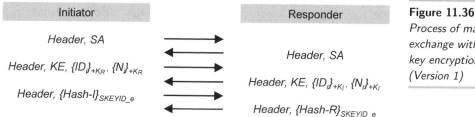

Figure 11.36
Process of main mode exchange with public key encryption (Version 1)

{m}+K_I denotes that *m* has been encrypted with public key *+K_I*

The first authentication method using public key encryption with main mode exchange is illustrated in Figure 11.36. The notation $\{m\}_{+K_I}$ denotes encryption of message element m with public key $+K_I$ of entity I. This method proves authenticity by requiring an entity to have knowledge of private key $-K_R$ or $-K_I$ that is needed to decrypt random number N_i or N_r. Computation of the master key SKEYID and consequently the hash values Hash-R and Hash-I is not possible without knowledge of the two random numbers. This means that an attacker has no possibility of successfully participating in a complete protocol run using eavesdropped or modified messages.

Authentication with public key encryption

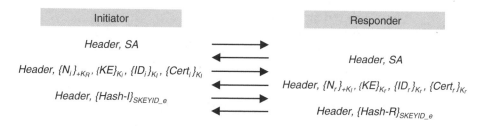

The second method of authentication using public key encryption as part of a main mode exchange is illustrated in Figure 11.37. The notation $\{m\}_{K_i}$ and $\{m\}_{K_r}$ denotes use of secret key K_i or K_r for the encryption of message element m.

These keys are computed as follows by initiator and responder:

$$K_i := H(N_i, CKY\text{-}I)$$
$$K_r := H(N_r, CKY\text{-}R)$$

The required random numbers N_i and N_r are not transmitted in plaintext but encrypted with the public key $+K_R$ or $+K_I$ and then

Figure 11.37
Process of main mode exchange with public key encryption (Version 2)

sent. Consequently, an attacker cannot compute these keys without knowing the corresponding private key. Both initiator and responder therefore prove authenticity by including both values into the SKEYID computation that is then entered into HASH-I or HASH-R. Encryption with keys K_i and K_r protects the confidentiality of the identity of both entities. The reason for using a symmetrical encryption scheme for this task is the higher processing speed that can be achieved with symmetric encryption.

Figure 11.38
Process of aggressive mode exchange with pre-shared key

Figure 11.38 shows an aggressive mode exchange with authentication using a pre-shared key. Because the identities of both the initiator and the responder have to be sent before a session key can be established between the entities, aggressive mode exchange is not able to protect the confidentiality of the identities from eavesdropping attackers. However, as it only needs half as many messages as compared with main mode exchange, this mode is normally used in cases where identity confidentiality is not important or cannot be protected anyway.

Other modes Similar modes exist for authentication based on digital signatures and public keys but will not be discussed here because they do not involve any new aspects.

11.8.2 Negotiation of IPSec-SA

After an ISAKMP-SA is established between two ISAKMP entities as the result of a Phase 1 exchange, it can be used to negotiate specific IPSec security associations. An ISAKMP-SA allows multiple IPSec associations to be negotiated simultaneously. Therefore, in principle both peer entities must be able to allocate individual ISAKMP messages to the respective SA negotiations and to synchronise the cryptographic state needed (e.g. initialisation vectors for the encryption of ISAKMP payloads).

Compared with the diverse combinations available with the different authentication methods using main mode and aggressive mode in Phase 1 exchanges, the specification for negotiating an IPSec-SA with IKE is relatively narrow. Figure 11.39 illustrates

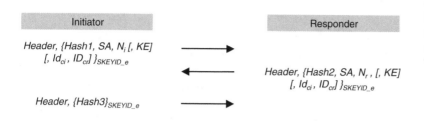

Figure 11.39
Process of quick mode exchange

the process of such a negotiation that is always executed in *quick mode*. The hash values used in quick mode exchange are defined as follows:

$$Hash1 := H(SKEYID\_a, M{-}ID, SA, N_i, [, KE][, ID_{ci}, ID_{cr}])$$
$$Hash2 := H(SKEYID\_a, M{-}ID, N_i, SA, N_r, [, KE][, ID_{ci}, ID_{cr}])$$
$$Hash3 := H(SKEYID\_a, 0, M{-}ID, N_i, N_r)$$

M–ID denotes the message identification contained in the ISAKMP message ID protocol field and used to allocate a message to a specific protocol run. This identification and the cookies in the ISAKMP protocol header identify the state belonging to the protocol run. Among other things, the state enables the calculation of the initialisation vector used to encrypt the payload.

The optional inclusion of identities ID_{ci} and ID_{cr} allows ISAKMP entities to negotiate security associations on behalf of other clients ('gateway scenario'). The optional key exchange payloads enable a new Diffie–Hellman exchange to be performed if the property of perfect forward secrecy is necessary for the IPSec-SA being established (see also Definition 7.4).

Delegated SA negotiation

The keying material for deriving session keys is computed according to the following rule by both entities upon successful completion of the exchange:

$$SessionKeyMaterial := H(SKEYID\_d[, g^{x \cdot y}], protocol, SPI, N_i, N_r)$$

11.9 Other Aspects of IPSec

We have learned that the ISAKMP and IKE protocols provide flexibility for the negotiation of security associations and that the IPSec architecture offers numerous options (transport or tunnel mode, AH, ESP, diverse supported cryptographic algorithms, etc.). On this basis the previous sections have shown how this architecture can be used to secure specific IP data streams. This discussion is now supplemented with some other aspects.

Compression in the context of IPSec

Note that if transmitted data packets are encrypted, the encrypted data usually cannot be compressed any further. This means that the desired reduction in data volume may not be achieved if compression is subsequently performed on the IP packets in the link layer (e.g. when connection is via a modem). It may therefore be necessary for the data to be compressed before encryption via the ESP protocol.

To deal with situations where compression of the IP packet payload is required, IETF specified the *IP Payload Compression Protocol (IPComp)* [SMPT01]. This protocol can be used in conjunction with IPSec. Its use can be defined within the framework of IPSec policies and it can also be proposed and negotiated for proposal payloads with IKE-agreed security associations.

Interoperability with protocol functions in gateway systems

Another aspect of IPSec is the interoperability with protocol functions in gateway systems. Use of end-to-end security measures in particular can lead to a range of problems with the evaluation or partial modification of protocol headers in gateways. Two basic examples that illustrate this problem are:

Firewalls

❑ *Interoperability problems with firewalls:* The main problem in connection with Internet firewalls (see also Chapter 13) is that end-to-end encryption of IP packets prevents a network operator from performing an authorised and often urgent inspection of the protocol headers.

Network address translation

❑ *Interoperability problems with network address translation (NAT):* Data packets encrypted in tunnel mode permit neither an analysis nor a modification of the address information in encapsulated IP packets. Even in transport mode there is a problem because the port addresses contained in Layer 4 protocol headers cannot simply be inspected for NAT. Furthermore, a receiver discards packets protected with AH when a NAT system makes changes to the outer IP protocol header necessary for address conversion. The packets would be discarded because their cryptographic checksums can no longer be correct in this case.

11.10　Summary

IPSec security services

IPSec is the security architecture for the Internet protocol developed by the Internet Engineering Task Force. It provides the following security services for IP packets:

❏ *Data origin authentication* and *data integrity* in connection with *replay protection*

❏ *Confidentiality* of payload or entire IP packets (limited protection against traffic flow analyses)

Securing IP packets exchanged between two systems requires that both systems define the methods and parameters to be used in the form of two *security associations (SA)* (one SA per direction). These associations also store the cryptographic context (session key, replay counter, etc.) for securing the packets.

Security associations

IPSec can be implemented in hosts and in gateways. Depending on whether the *cryptographic end points* of a security association coincide with the *communication end points* of the protected data stream, the IP packets concerned are secured in either *transport* or in *tunnel mode*. The latter mode performs a complete encapsulation of the IP packets.

Transport and tunnel modes

IPSec defines two fundamental security protocols:

❏ *Authentication header (AH)*, which ensures the authenticity and freshness of IP packets including the immutable parts of their outer protocol header.

Authentication header

❏ *Encapsulating security payload (ESP)*, which optionally allows the encryption and authentication of the payload but cannot protect the authenticity of the immutable fields of the outer protocol header.

Encapsulating security payload

The two mutually complementary protocols *Internet Security Association and Key Management Protocol (ISAKMP)* and *Internet Key Exchange (IKE)* were defined for the negotiation of security associations in connection with entity authentication. ISKAMP specifies fundamental data formats and protocol procedures whereas IKE defines specific authentication protocols.

Entity authentication with ISAKMP and IKE

11.11 Supplemental Reading

[HC98]   HARKINS, D.; CARREL, D.:   *The Internet Key Exchange (IKE)*. November 1998. – RFC 2409, IETF, Status: Proposed standard, `ftp://ftp.internic.net/rfc/rfc2409.txt`

[KA98a]   KENT, S.; ATKINSON, R.: *IP Authentication Header*. November 1998. – RFC 2402, IETF, Status: Proposed standard, `ftp://ftp.internic.net/rfc/rfc2402.txt`

[KA98b]   KENT, S.; ATKINSON, R.: *IP Encapsulating Security Payload (ESP)*. November 1998. – RFC 2406, IETF, Status: Proposed standard, `ftp://ftp.internic.net/rfc/rfc2406.txt`

[KA98c]   KENT, S.; ATKINSON, R.: *Security Architecture for the Internet Protocol*. November 1998. – RFC 2401, IETF, Status: Proposed standard, `ftp://ftp.internic.net/rfc/rfc2401.txt`

[KP00]   KEROMYTIS, A.; PROVOS, N.: *The Use of HMAC-RIPEMD-160-96 within ESP and AH*. June 2000. – RFC 2857, IETF, Status: Proposed standard, `ftp://ftp.internic.net/rfc/rfc2857.txt`

[MD98]   MADSON, C.; DORASWAMY, N.: *The ESP DES-CBC Cipher Algorithm With Explicit IV*. November 1998. – RFC 2405, IETF, Status: Proposed standard, `ftp://ftp.internic.net/rfc/rfc2405.txt`

[MG98a]   MADSON, C.; GLENN, R.: *The Use of HMAC-MD5-96 within ESP and AH*. November 1998. – RFC 2403, IETF, Status: Proposed standard, `ftp://ftp.internic.net/rfc/rfc2403.txt`

[MG98b]   MADSON, C.; GLENN, R.: *The Use of HMAC-SHA-1-96 within ESP and AH*. November 1998. – RFC 2404, IETF, Status: Proposed standard, `ftp://ftp.internic.net/rfc/rfc2404.txt`

[MSST98]   MAUGHAN, D.; SCHERTLER, M.; SCHNEIDER, M.; TURNER, J.: *Internet Security Association and Key Management Protocol (ISAKMP)*. November 1998. – RFC 2408, IETF, Status: Proposed standard, `ftp://ftp.internic.net/rfc/rfc2408.txt`

[Pip98]   PIPER, D.: *The Internet IP Security Domain of Interpretation for ISAKMP*. November 1998. – RFC 2407, IETF, Status: Proposed standard, `ftp://ftp.internic.net/rfc/rfc2407.txt`

[SMPT01]   SHACHAM, A.; MONSOUR, B.; PEREIRA, R.; THOMAS, M.: *IP Payload Compression Protocol (IPComp)*. September 2001. – RFC 3173, IETF, Status: Proposed standard, `ftp://ftp.internic.net/rfc/rfc3173.txt`

11.12 Questions

1. Why are the principle of connectionless data transmission and the requirements for secure communication not compatible with one another, and which temporary measure is used in the IPSec security architecture to compensate for this shortcoming?

2. Would it be possible to implement secure data transmission more efficiently in a connection-oriented communication architecture and, if yes, in which respect?

3. Why is the authentication header not able to protect all fields of the outer IP packet header?

4. Explain the tasks of the two conceptual databases, security association database and security policy database.

5. Why is the security parameter index for a security association in principle allocated by the receiving system?

6. Compare the task and function of the sliding window method for the correction of transmission errors with IPSec replay protection.

7. What are the uses of transport and tunnel mode and how do they differ?

8. Can you give a reason why it can be useful to operate an ESP-SA in tunnel mode although the cryptographic end points coincide with the communication end points?

9. What are the main disadvantages of nested IPSec security associations in terms of efficiency?

10. Can IPSec also be used to implement access control functions and, if yes, how?

11. Why does the checksum of the outer IP protocol header have to be recomputed when an ESP or an AH protocol header is created?

12. Why does it make sense when processing incoming IPSec packets to verify that the sequence numbers are acceptable before continuing with other cryptographic tests?

13. Is it possible that an IP packet has to run through IPSec protocol processing multiple times in one and the same system?

14. Why is the authenticity of an AH or an ESP packet always verified before the replay window is advanced?

15. Explain the protective effects of the ISAKMP cookie mechanism.

16. Explain how the logical functions 'AND' and 'OR' are realised when a protection suite is negotiated with ISAKMP.

17. Compare the four modes of ISAKMP main mode with respect to the complexity created through cryptographic operations.

18. Is compressed data transmission using the IP payload compression protocol combined with ESP disadvantageous compared to compression directly performed via a modem?

12 Transport Layer Security Protocols

Unlike the network layer, which enables communication between hosts, the transport layer implements communication between application processes. Its main tasks include:

Transport layer tasks

❑ Isolation of higher-order protocol layers from the technology, structure and deficiencies of the deployed communication technology.

❑ Transparent transmission of user data.

❑ Global addressing of application processes independently of the addressing formats of the lower communication layers.

The overall goal is to provide an efficient and reliable transmission service.

The security protocols of the transport layer enhance the service of the transport layer by assuring additional security properties. As security protocols usually require and are run on a reliable transport service, they actually represent *session layer protocols* in accordance with the OSI model for open communication systems.

However, because the OSI model has not been *'en vogue'* since around the mid-1990s, the session layer protocols are usually referred to as *transport layer security protocols*. This chapter deals with the most common protocols in this category: *Secure Socket Layer (SSL)*, the protocol derived from it *Transport Layer Security (TLS)* and the independent protocol *Secure Shell (SSH)*.

12.1 Secure Socket Layer (SSL)

The *Secure Socket Layer (SSL)* protocol was originally designed with the primary goal of protecting sessions of the *Hypertext Transfer Protocol (HTTP)*. In the early 1990s, a similar competing approach existed called the *Secure HTTP (S-HTTP)* protocol. However, as S-HTTP-capable web browsers were not free of charge, whereas SSL Version 2.0 was included at no additional cost in the Netscape Communications browsers, SSL quickly became the dominant security protocol for HTTP.

S-HTTP was a competitor to SSL

PCT was another
competitor

Version 2.0 of the protocol contained a number of deficiencies, so Microsoft developed another competitive protocol called *Private Communication Technology (PCT)*. However, Microsoft's web browsers were not very established at the time and so the improved SSL Version 3.0 [FKK96] from Netscape Communications managed to hold its ground as the standard protocol for securing HTTP traffic.

SSL was also
standardised as TLS

Despite its origins as a security protocol for HTTP, SSL can be used to secure any application run over the transport protocol TCP. In 1996 the Internet Engineering Task Force therefore decided to develop a generic *Transport Layer Security (TLS)* protocol based on SSL.

12.1.1 Security Services and Protocol Architecture

SSL Version 3.0 provides the following security services:

❑ *Entity authentication:* Prior to any communication between client and server, an authentication exchange is performed to verify the identity of the peer entity either only to the client or also to the server. After successful authentication, an *SSL session* is established between the two entities.

❑ *Confidentiality of user data:* If agreed during negotiation of the SSL session, the user data is encrypted. SSL offers a range of algorithms for this purpose, e.g. RC4, DES, 3DES and IDEA, the use of which can be negotiated during session establishment.

❑ *Data origin authentication and data integrity:* Each message is secured with a MAC that is computed using a cryptographic hash function. SSL Version 3.0 originally used *prefix–suffix mode* for this purpose although there were some security concerns about it. Either MD5 or SHA can be negotiated as the underlying cryptographic hash function.

Session and
connection state

SSL uses the concept of *sessions* that was already anchored in the OSI model. Prior to actual communication, client and server establish a session in which the parameters for securing the communication are negotiated. As in the OSI model, a session can run over multiple transport layer connections if it is negotiated as a 'resumable' session. SSL makes a clear distinction between the following two state sets:

Session state

❑ The *session state* stores the following data:

- *Session identifier:* a byte sequence chosen by the server to identify the session.

- *Peer certificate:* an (optional) X.509v3 certificate from the peer entity.

- *Compression method:* denotes the data compression algorithm used before encryption.

- *Cipher spec:* specifies the cryptographic algorithms and their parameters.

- *Master secret:* a 48-byte long secret negotiated between client and server.

- *Is resumable:* indicates whether the session can be resumed or duplicated (i.e., multiple transport connections supported simultaneously).

- In *connection state* the following information is provided over the current transport layer connection(s):   *Connection state*

  - *Server and client random:* contains random byte sequences chosen by the server or the client.

  - *Server write MAC secret:* is included by the server in its MAC computations.

  - *Client write MAC secret:* the same as above but for client messages.

  - *Server write key:* the key for encrypting and decrypting messages from the server.

  - *Client write key:* the same as above but for messages from the client.

The SSL protocol itself is structured as a layered and modular protocol architecture as shown in Figure 12.1.

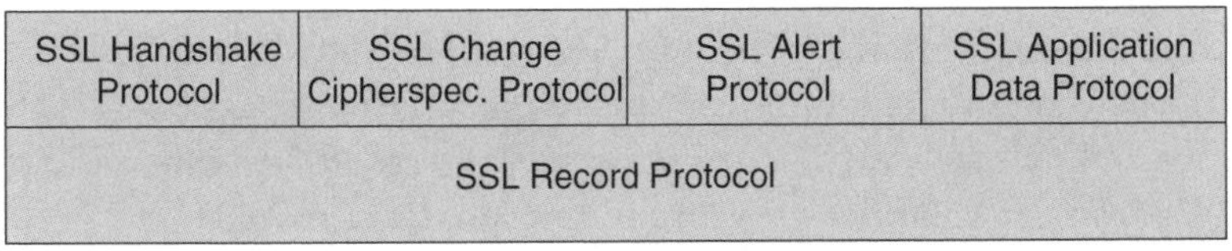

SSL Handshake Protocol	SSL Change Cipherspec. Protocol	SSL Alert Protocol	SSL Application Data Protocol
SSL Record Protocol			

Figure 12.1
Architecture of the secure socket layer protocol

The *Record protocol* is used as the basis for data exchange in SSL   *Record protocol*
sessions. Its tasks include the fragmentation of user data into

plaintext blocks (called 'records') no longer than 2^{14} octets, the optional compression of plaintext blocks and the optional encryption and authentication of plaintext blocks.

Handshake protocol

Change cipherspec. protocol

Alert protocol

Application data protocol

The *handshake protocol* is used for entity authentication and session negotiation. The *change cipherspec. protocol* signals changes in certain cryptographic parameters (session keys, algorithms, etc.) and the *Alert protocol* signals error conditions. The *application data protocol* is a transparent interface to the record protocol and is used by applications for data unit exchanges with an SSL protocol entity.

12.1.2   The Record Protocol

Figure 12.2 illustrates how an SSL record is structured. The *(content) type* field identifies the SSL protocol contained in the record: change cipherspec (20), alert (21), handshake (22) or application data (23). The *version* identifies the SSL protocol version (major = 3, minor = 0) and the *length* field contains the length of user data in octets. The maximum user data length is $2^{14} + 2^{10}$.

Record protocol processing procedures

The protocol of the sending entity is processed according to the following procedure: First the user data (or the data of the handshake protocol, change cipherspec. protocol or alert protocol) is fragmented into frames of a maximum length of 2^{14} octets. More than one message of the same protocol type can be assembled in one frame.

Figure 12.2
Frame format of the SSL record layer protocol

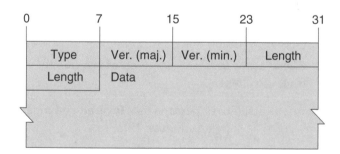

Compression

After fragmentation the frames are compressed. Although compression is not the default, it can be negotiated at session establishment. However, the data is only compressed for transmission if the compression does not produce a length of more than 2^{10}. (A lengthening can occur for example if the data was already compressed, e.g. as with a JPEG file).

MAC computation

After compression a MAC is computed ('||' denotes concatenation):

$$MAC := H\big(WriteMacSecret \parallel pad_2 \parallel$$
$$H(WriteMacSecret \parallel pad_1 \parallel seqnum \parallel$$
$$length \parallel data)\big)$$

The sequence number included in the MAC computation is not transmitted because it is implicitly known to the receiving side. The reliable TCP protocol is used to implement the underlying transport layer connection so that any loss of data units is detected.

The compressed data with the MAC is then encrypted with the method negotiated at session establishment. Depending on the method used, the data first has to be expanded to a multiple of a specific block length. The data is then sent to the service interface of the transport protocol. *Encryption*

On the receiving side the data is successively decrypted, its authenticity is verified and it is decompressed, defragmented and subsequently delivered to the appropriate application or the SSL higher layer protocol. *Processing at the receiver*

12.1.3 The Handshake Protocol

The SSL handshake protocol is used to negotiate SSL sessions. As mentioned earlier, a session can be negotiated so that it can be duplicated or resumed at a later time, thus allowing an established cryptographic context to be reused. This feature is particularly important for the efficiency of secured HTTP communication since each unit of a hypertext document (e.g. an HTML document with references to graphics, sound, etc.) is normally transported in a separate TCP connection. If a separate complete authentication protocol with flexible negotiation of cryptographic parameters were required each time, this would have a negative effect on data throughput for the client. Session resumption on the other hand allows the use of shorter exchanges. *Negotiation of SSL sessions* *Resumption of sessions*

Figure 12.3 presents an overview of a detailed exchange for negotiating an SSL session. The first message contains a *ClientHello*, which combines the protocol version used by the client, a random number, possibly a session ID, a list of *cipher suites* supported by the client and a list of allowable compression methods. *Protocol run* *First message from client*

The server responds to this message with a *ServerHello*, the optional message elements *ServerCertificate*, *CertificateRequest* *First message from server*

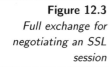

Figure 12.3
Full exchange for negotiating an SSL session

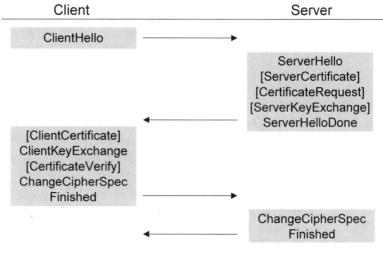

[...] denotes optional message elements

and *ServerKeyExchange* as well as a *ServerHelloDone* that concludes the response. The *ServerHello* combines the protocol version supported by it, a random number, possibly a session ID along with the *cipher suite* selected by the server for the session and a designated compression method.

If the session ID contains the value 0, the server is signalling that it is not storing the connection and therefore the connection can neither be duplicated nor resumed at a later time. A value other than 0 means that the client can resume the session at a later time (see below). The certificate included by the server enables the client to verify the identity of the server. The server is usually not interested in the identity of the client, e.g. if it is merely providing information to a client. However, certain applications also require an authentication of the client. In such cases, the server also requests a certificate from the client. Depending on the method used to negotiate the session keys, the server also sends information for the key negotiation, e.g. its public Diffie–Hellman values or a public RSA key. The *ServerHelloDone* contains no further information and is only used to signal that the server is now waiting for a response from the client. This message is necessary because the client otherwise has no way of knowing whether the server is still going to send an optional message ele-

The server is usually not interested in the identity of the client

ment or whether its response should be regarded as being complete.

The client responds to the message from the server with its certificate (if required), a *ClientKeyExchange*, an optional *CertificateVerify* and the message elements *ChangeCipherSpec* and *Finished*. The *CertificateVerify* enables the client to prove that it has a private key that matches the public key certified in its certificate (provided this certificate was transmitted and certifies a key with signing capability). The client uses *ChangeCipherSpec* to signal that all subsequent message elements it sends are protected with the negotiated cryptographic protection suite. The first message element protected in this way is therefore the element *Finished*, sent immediately after ChangeCipherSpec containing a hash value computed over the shared secret (see below) negotiated during the exchange and all message elements (including ClientHello) sent previously by the client.

Second message from client

The server responds to this message with a *ChangeCipherSpec* and a *Finished*. After the client and the server have both sent a Finished message and have verified the Finished message of their respective peer entity, they consider the session as established and can begin sending protected user data.

Second message from server

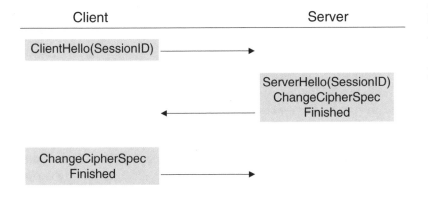

Figure 12.4
Abbreviated exchange for negotiating an SSL session

Figure 12.4 shows an abbreviated exchange for resuming or duplicating a session. In this case the client sends the session ID of the session being resumed in its ClientHello message. If the server has stored the session locally and is receptive to resuming the session, it includes the message elements ChangeCipherSpec and Finished immediately after its ServerHello response. The verification of authenticity in this case is solely based on the hash values in the Finished messages, which include the random numbers contained in

Resuming a session

the Hello messages and thus guarantee the freshness of the message exchanges. If the server is not receptive to resuming the session, it responds with the message elements shown in Figure 12.3, thereby initiating a full exchange between client and server.

Cipher suites   The exchanged *cipher suites* in the Hello messages always reference a combination of cryptographic schemes for establishing and protecting the subsequent session. The values defined for this purpose by the SSL specification are listed in Table 12.1.

12.1.4   Authentication and Negotiation of Session Keys

Negotiation of   SSL supports three methods for negotiating a *pre-master secret*
pre-master secret   from which other session keys are derived:

RSA-protected   ❏ *RSA-protected negotiation:* With this method a pre-master
negotiation   secret is randomly created by the client, encrypted using the
                         public key of the server, and sent to the server. The server
                         does not send a separate KeyExchange message element of
                         its own to the client as it is not actively involved in creating
                         the shared secret.

Diffie–Hellman   ❏ *Diffie–Hellman:* This involves executing a conventional
                         Diffie–Hellman key negotiation protocol with the pre-master
                         secret derived from the shared secret $g^{x \cdot y} \mod p$.

Fortezza   ❏ *Fortezza:* This is a non-published method that was developed
                         by the National Security Agency and supports key escrow for
                         government agencies. As there are justified security concerns
                         about this type of method, we will not cover it in this book.

Securing HTTP   As mentioned at the beginning of this chapter, SSL was originally
traffic   designed to secure HTTP traffic. Therefore, a client usually wants
                      to access a web server and have assurance of its authenticity. Consequently, the web server sends a certificate of its public key right
                      after its ServerHello message. This certificate can optionally contain the public Diffie–Hellman values of the server (cyclical group
                      and primitive root of group, also see Section 4.5). In this case
                      the client must support the same group and primitive root, which
                      can become a problem if the client has also had its public Diffie–
                      Hellman values certified and they turn out to be based on a different cyclical group. Alternatively, the server can also send its public
                      Diffie–Hellman values in a KeyExchange message element to the

Coding	Symbolic description
{0x00,0x00}	SSL NULL With NULL NULL
{0x00,0x01}	SSL RSA With NULL MD5
{0x00,0x02}	SSL RSA With NULL SHA
{0x00,0x03}	SSL RSA Export With RC4 40 MD5
{0x00,0x04}	SSL RSA With RC4 128 MD5
{0x00,0x05}	SSL RSA With RC4 128 SHA
{0x00,0x06}	SSL RSA Export With RC2 CBC 40 MD5
{0x00,0x07}	SSL RSA With IDEA CBC SHA
{0x00,0x08}	SSL RSA Export With DES40 CBC SHA
{0x00,0x09}	SSL RSA With DES CBC SHA
{0x00,0x0A}	SSL RSA With 3DES EDE CBC SHA
{0x00,0x0B}	SSL DH DSS Export With DES40 CBC SHA
{0x00,0x0C}	SSL DH DSS With DES CBC SHA
{0x00,0x0D}	SSL DH DSS With 3DES EDE CBC SHA
{0x00,0x0E}	SSL DH RSA Export With DES40 CBC SHA
{0x00,0x0F}	SSL DH RSA With DES CBC SHA
{0x00,0x10}	SSL DH RSA With 3DES EDE CBC SHA
{0x00,0x11}	SSL DHE DSS Export With DES40 CBC SHA
{0x00,0x12}	SSL DHE DSS With DES CBC SHA
{0x00,0x13}	SSL DHE DSS With 3DES EDE CBC SHA
{0x00,0x14}	SSL DHE RSA Export With DES40 CBC SHA
{0x00,0x15}	SSL DHE RSA With DES CBC SHA
{0x00,0x16}	SSL DHE RSA With 3DES EDE CBC SHA
{0x00,0x17}	SSL DH anon Export With RC4 40 MD5
{0x00,0x18}	SSL DH anon With RC4 128 MD5
{0x00,0x19}	SSL DH anon Export With DES40 CBC SHA
{0x00,0x1A}	SSL DH anon With DES CBC SHA
{0x00,0x1B}	SSL DH anon With 3DES EDE CBC SHA
{0x00,0x1C}	SSL Fortezza DMS With NULL SHA
{0x00,0x1D}	SSL Fortezza DMS With Fortezza CBC SHA

Table 12.1
SSL Version 3.0
Cipher Suites

client. In each of the cases mentioned, the client verifies the certificate of the server and uses it either for RSA, Diffie–Hellman or Fortezza-based key negotiation to establish the pre-master secret.

Computation of pre-master secret

As the next step, the pre-master secret together with the two random numbers of the ClientHello and ServerHello messages are used to derive the 48-byte long *master secret* according to the following rule (the character '$\|$' denotes a concatenation of byte sequences):

$$master\text{–}secret := MD5\big(pre\text{–}master\text{–}secret \;\|$$
$$SHA(\,'A' \; \| \; pre\text{–}master\text{–}secret \;\|$$
$$ClientHello.random \; \| \; ServerHello.random)\big) \;\|$$
$$MD5\big(pre\text{–}master\text{–}secret \;\|$$
$$SHA(\,'BB' \; \| \; pre\text{–}master\text{–}secret \;\|$$
$$ClientHello.random \; \| \; ServerHello.random)\big) \;\|$$
$$MD5\big(pre\text{–}master\text{–}secret \;\|$$
$$SHA(\,'CCC' \; \| \; pre\text{–}master\text{–}secret \;\|$$
$$ClientHello.random \; \| \; ServerHello.random)\big)$$

The use of the two hash functions MD5 and SHA was defined with the intention that the scheme should still be secure even if cryptographic deficiencies are discovered with one of the two hash functions.

Derivation of session keys

A sufficiently long *key block* is derived from the master secret and the random numbers of the two Hello messages. The necessary session keys and initialisation vectors are extracted from the master secret without overlapping.

A key block is constructed using the following computation rule:

$$kb := MD5\big(master\text{–}secret \; \| \; SHA(\,'A' \; \| \; master\text{–}secret \;\|$$
$$ClientHello.random \; \| \; ServerHello.random)\big) \;\|$$
$$MD5\big(master\text{–}secret \; \| \; SHA(\,'BB' \; \| \; master\text{–}secret \;\|$$
$$ClientHello.random \; \| \; ServerHello.random)\big) \;\|$$
$$[\dots]$$

The following session keys and initialisation vectors are then extracted from this:

$$ClientWriteMacSecret := kb[1, CipherSpec.HashSize]$$
$$ServerWriteMacSecret := kb[i_1, i_1 + CipherSpec.HashSize - 1]$$
$$ClientWriteKey := kb[i_2, i_2 + CipherSpec.KeySize - 1]$$
$$ServerWriteKey := kb[i_3, i_3 + CipherSpec.KeySize - 1]$$
$$ClientWriteIV := kb[i_4, i_4 + CipherSpec.IvSize - 1]$$
$$ServerWriteIV := kb[i_5, i_5 + CipherSpec.IvSize - 1]$$

The pre-master secret from which the session keys are derived is used to authenticate the server and possibly also the client.

Aspects of authentication

Using RSA-protected negotiation, the client encrypts the pre-master secret with the public key of the server. As the client can verify the authenticity of this key by evaluating the server certificate, it has the assurance that only the server is able to decrypt the pre-master key. When the client receives the server's Finished message, which contains a hash value over all exchanged message elements and the master secret derived from the pre-master secret, it can conclude that the server is authentic. However, this method does not allow the server to verify the authenticity of the client because any arbitrary client is able to send it a pre-master secret encrypted with the public key of the server. If verification of the authenticity of the client is also required, the client uses RSA-protected negotiation to send an additional client certificate and a CertificateVerify message, which contains a hash value (computed with MD5 or SHA) over the master secret and all previously sent message elements.

RSA-protected negotiation

If Diffie–Hellman key exchange with certified public Diffie–Hellman values is used, the authenticity of the server can be directly deduced from the negotiated master secret that is included in the hash value of the Finished message. In case the client also sends a certificate with public Diffie–Hellman values that match those of the server, its authenticity can be directly verified on the basis of its Finished message.

Diffie–Hellman key exchange

SSL also support anonymous key negotiation without authentication although in principle this version cannot offer protection against potential man-in-the-middle attacks. It is therefore advisable for it to be deactivated in the local configuration.

Anonymous key negotiation

12.1.5   A Shortcoming in the Handshake Protocol

In 1998, Bleichenbacher discovered a shortcoming in the encryption standard *PKCS #1* (Version 1.5), which is used in the SSL handshake protocol [BKS98]. Attackers who use RSA-protected negotiation to obtain a pre-master secret are able to exploit this shortcoming.

Formatting based on PKCS #1　　In this case the client formats the pre-master secret prior to encryption with the RSA algorithm using the standard PKCS #1, which formats and encrypts a message M as follows:

$$EM := 0x02 \parallel PS \parallel 0x00 \parallel M$$
$$C := RSA(+K_{Server}, EM)$$

PS above denotes a 'padding string' with a minimum of eight randomly generated octets that are always unequal to '0x00' and should add a random component to the plaintext and pad it to the length of the RSA module being used.

After the server deciphers the ciphertext C, it verifies whether the first octet of the plaintext is equal to '0x02' and whether the plaintext also contains an octet with the value equal to '0x00'. If the verification is unsuccessful, the server responds with an error message. It is precisely this error message that an attacker can use to execute an *Oracle attack)*.

Operation of an Oracle attack　　With this type of attack, the attacker Eve tries to discover the pre-master secret negotiated in a previously eavesdropped handshake exchange between Alice (client) and Bob (server) as well as all the session keys derived from it. So, Eve has eavesdropped on ciphertext C and wants to recover the corresponding plaintext EM.

She therefore generates a series of ciphertexts C_i, which she sends to Bob as part of a bogus SSL session establishment. These have a specific relationship to one another (e and n are the two values of Bob's public key):

$$C_i = C \cdot R_i^e \mod n$$

The R_i values are constructed in a way that depends on previous 'good' R_i values that were processed by Bob without generating an error message. This indicates they were decrypted to valid plaintexts using PKCS #1. From the information that specific C_i have resulted in valid 'PKCS #1' plaintexts, Eve can now deduce certain bits of the corresponding message $M_i \equiv C_i^d \equiv M \cdot R_i \mod n$.

From the inferred bits of $M \cdot R_i$ MOD n for a sufficient number of R_i, Eve is able to reduce the size of the interval containing message M. Each 'good' ciphertext halves the interval in question so that with a sufficient number of 'good' values R_i Eve is able to determine message M and, consequently, the pre-master secret agreed between Alice and Bob.

With Version 1.5 of PKCS #1 used in SSL Version 3.0, approximately one of 2^{16} to 2^{18} values R_i prove to be 'good'. On the basis of the currently common RSA key that has a 1024-bit module length, with this method Eve will require approximately 2^{20} ciphertexts and requests to Bob to recover a plaintext. Therefore, after approximately one million bogus handshake exchanges all of which are disrupted by Bob (because of the 'not good' R_i) or Eve (in case of 'good' R_i), Eve is able to discover the pre-master secret previously negotiated between Alice and Bob and all the session keys derived from it.

With a 1024-bit module length an attacker only needs around 2^{20} requests to compute a plaintext!

In summary, this type of attack shows that *subtle protocol interactions (here: between PKCS #1 and SSL) can result in the failure of a protocol even if the basic cryptographic algorithm (here: RSA) is not broken itself.*

Various countermeasures exist to combat this type of attack. A naïve but effective measure is to change the asymmetric keys regularly so that attackers are unable to try out sufficient C_i to determine the source text that belongs to C. However, a considerable effort is required as the public key of the server must be certified to enable the client to verify its authenticity.

Countermeasures

Another approach is to reduce the probability of an attacker finding 'good' ciphertexts. This can be done by carefully checking the format of received plaintexts and showing identical behaviour (error notification, time behaviour, etc.) to prevent the client from misusing the server as an Oracle. Another recommendation is to require that the client can prove it knows the plaintext belonging to a ciphertext before it receives information as to whether the ciphertext could be decrypted into a valid plaintext. One possibility for implementing these two ideas is the addition of a strict structure to plaintexts, for example through the concatenation of a cryptographic hash value on each plaintext message. However, it is important that no weaknesses are introduced that can be exploited for another category of attacks [CFPR96].

The client should prove it knows the encrypted plaintext

The best countermeasure to protect against Oracle attacks on protocols using the RSA algorithm is to change the asymmetrical encryption protocol, in this case PKCS #1. Version 2.1 of the standard provides for the plaintext to be prepared using a method called *Optimal Asymmetric Encryption Padding (OAEP)* prior to actual encryption. The aim of this type of preparation is to prevent attackers from producing valid ciphertexts if they do not know the corresponding plaintexts. This property of the modified 'PKCS #1' protocol is called *'plaintext aware'*.

12.2   Transport Layer Security (TLS)

In 1996 the IETF founded a working group to specify a security protocol called *Transport Layer Security (TLS)*. Officially, the protocols SSL, SSH and PCT were to be used as the common input, but it was quickly decided that the protocol SSL Version 3.0 with the following modifications should be adopted as TLS:

TLS is based on SSL Version 3.0

❏ The prefix–suffix construction originally used in SSL Version 3.0 to compute cryptographic hash values should be replaced by the HMAC construction.

❏ The cipher suites based on Fortezza should be removed from the protocol because they are based on a non-published technology.

❏ The handshake protocol should be enhanced to include an authentication version based on the Digital Signature Standard (DSS).

❏ The protocol specification should be modularised and the record and handshake protocols in particular specified in separate documents.

The last requirement was not actually implemented so the first draft of the TLS specification largely resembles SSL Version 3.0. This also applies to RFC 2246 that was adopted in 1999 [DA99].

Adaptations to obtain global export permission

Several cipher suites stipulating the use of keys with a maximum entropy of 40 bits were defined in order to obtain global export permission for TLS-conformant products. These cipher suites contain the word 'Export' in their symbolic labels. They actually provide no protection against attackers since brute-force attacks can be executed today to 40-bit long keys on conventional computers with adequate processing speeds. As a result of the change in US export policy regarding cryptography, these cipher suites have lost their significance.

Supported cryptographic algorithms and protocols

In terms of its protocol functions, TLS is no different than SSL. The protocol supports the following cryptographic algorithms and protocols in its cipher suites:

❏ Authentication and key exchange: Diffie–Hellman key exchange with or without digital signatures (DSS or RSA),

Diffie–Hellman key exchange with certified public values, RSA-protected negotiation;

❏ encryption algorithms: IDEA, DES, 3DES, RC2 always in CBC mode, RC4, no encryption;

❏ cryptographic hash functions: MD5, SHA, without hash values.

12.3   Secure Shell (SSH)

The protocol *Secure Shell (SSH)* was originally developed by Ylönen at Helsinki University in Finland. As the author also provided a free implementation with source code for general use, the protocol was widely used in the Internet. Although the author subsequently commercialised the development of SSH, free implementations of the current protocol version still exist with *OpenSSH* being the one most widely used.

Version 2.0 of the SSH specification was submitted to the IETF in 1997 and since then has been refined on a regular basis in a series of updated Internet drafts [YKS+01a, YKS+01b, YKS+01c, YKS+01d].

Standardisation in IETF

SSH was designed with the goal of creating a secure replacement for the *R-Tools (rlogin, rsh, rcp, rdist)* in the *Unix* operating system. It thus represents an application layer or session layer protocol. However, as it also contains a generic security protocol for transport connections and supports the encapsulation of transport connections (tunnelling), it is referred to in this chapter as a transport layer security protocol.

Goal of SSH

Version 2 of the SSH specification is divided into four documents:

❏ SSH Protocol Architecture [YKS+01c]

❏ SSH Transport Layer Protocol [YKS+01d]

❏ SSH Authentication Protocol [YKS+01a]

❏ SSH Connection Protocol [YKS+01b]

The protocol architecture is based on a client/server approach. Each server has at least one public key that is used for the entire computer. SSH supports two fundamental trust models: With the simple model, each client has access to a local database that stores

Protocol architecture

Trust models

the public key for each computer known to it. SSH also supports the allocation of public keys to servers on the basis of certificates. In this case clients know the public key(s) of one or more certification entities that certify the public keys of the servers. In addition, the protocol enables a flexible negotiation of the protocol mechanisms used in a session, including encryption, data integrity, key negotiation and compression, along with the respective algorithms and parameters.

Negotiation of protocol mechanisms

12.3.1   SSH Transport Protocol

SSH security services

The SSH transport protocol runs on top of a reliable transport protocol (usually TCP) to provide the following security services for data exchanged in a session:

❏ server authentication (host-related);

❏ user data encryption (after optional prior compression);

❏ data origin authentication.

Cryptographic algorithms and protocols

The server is authenticated using RSA or DSS signatures. A session key is usually agreed with authenticated Diffie–Hellman exchange at the same time. The algorithms 3DES, Blowfish, Twofish, AES, Serpent, IDEA and CAST in cipher block chaining mode and RC4 are supported for the encryption of the user data. For data integrity authentication the hash values are computed alternatively with MD5 or SHA using the HMAC construction.

Protocol fields

Figure 12.5 shows the frame format of the SSH transport protocol. The significance of the protocol fields contained in the frame is as follows:

Packet length

❏ *Packet length* denotes the length of the frame minus this field.

Pad length

❏ *Pad length* indicates how many octets were used to lengthen the packet before encryption and must contain a value between 4 and 255.

Payload

❏ *Payload* contains the actual payload of the frame. If encryption of the user data was negotiated at the time of session establishment, the payload is encrypted after optional compression and computation of the MAC.

Padding

❏ *Padding* denotes randomly selected octets that are used to pad the payload to an integer multiple of 8 or the block

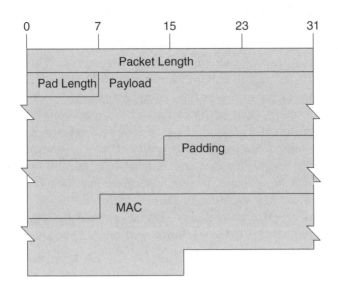

Figure 12.5
Frame format of SSH transport protocol

length of the encryption scheme (depending on which value is higher).

❏ *MAC* contains the optional authentication value of the message that is computed according to the following rule: *MAC*

$$MAC := HMAC(SharedSecret, SeqNum \parallel UnencryptedPacket)$$

SeqNum denotes a 32-bit sequence number incremented for each frame and *UnencryptedPacket* denotes the entire frame without the MAC field.

Like the SSL record protocol, the frames with the SSH transport protocol are not aligned to 32-bit boundaries. Such an alignment is particularly important for protocols up to the network layer because it usually enables protocols to be processed efficiently in gateway systems. As protocol data units above the transport layer are no longer processed in gateway systems, this is something that can be eliminated without impacting performance.   *No alignment to 32-bit boundaries*

12.3.2   Parameter Negotiation and Server Authentication

When an SSH transport protocol connection is established, client and server agree on the cryptographic scheme that should be used.

Negotiation of supported scheme and preferences

Each peer entity sends packet referred to as '*Kexinit*' with a specification of supported schemes in the order of local preference. Both entities then iterate over the list of the client and select the first scheme that is also supported by the server. This enables schemes to be negotiated for server authentication, encryption, MAC computation and user data compression.

Key negotiation

Furthermore, each entity can attach a *key exchange packet* for the negotiation of a session key by selecting a scheme it assumes also to be supported by the peer entity. If this assumption turns out to be false, the key exchange packet is discarded by the peer entity and a new packet is sent in accordance with the scheme negotiated for key exchange. This situation does not occur frequently in practice as the only scheme defined for key negotiation in [YKS+01d] is Diffie–Hellman with SHA-1 and the following group:

$$p := 2^{1024} - 2^{960} - 1 + (2^{64} \cdot \lfloor 2^{894} \cdot \pi + 129093 \rfloor)$$

$$g := 2$$

The order of this group is $(p-1)/2$. If the peer entity supports the scheme used in the key exchange packet, it processes the packet as the first packet for the key exchange. This saves the full roundtrip time between client and server.

Protocol run

The scheme for key negotiation and server authentication specified in [YKS+01d] with the group given above comprises the following protocol procedures:

1. The client selects a random number x, computes $e := g^x \text{ MOD } p$ and sends e to the server.

2. The server selects a random number y, computes $f := g^y \text{ MOD } p$ and sends f to the client. The server does not have to wait for the first packet from the client with the value e and instead can execute this operation in parallel with the first protocol step of the client.

3. Upon receipt of e the server also computes:
   $K := e^y \text{ MOD } p$ as well as the hash value
   $$h := Hash(ver_C, ver_S, kexinit_C, kexinit_S, +K_S, e, f, K)$$

The message elements *ver* and *kexinit* denote the protocol version and the initial message for the algorithm negotiation of the client or the server. The server signs the hash value h with its private key $-K_S$, i.e., it computes $s := E(-K_S, h)$, and sends the message $(+K_S, f, s)$ to the client.

4. Upon receipt of this message the client verifies the authenticity of the public key $+K_S$ (either by querying its local database or by verifying the key certificate belonging to $+K_S$). It then computes $K := f^x$ MOD p and the hash value h. It also verifies the signature over the hash value h. If the verification is successful, the client is assured that it has negotiated a shared secret K with the server in possession of key $-K_S$.

However, the server cannot use this exchange to deduce the authenticity of the client; the SSH authentication protocol described in the next section is used for this purpose.

The following session keys ($EK \approx$ encryption key, $IK \approx$ integrity key) are then derived by both sides from the negotiated secret K and hash value h, with hash value h of the initial key exchange also used as the session ID:

Derivation of session keys

$$IV_{Client2Server} := Hash(K, h, 'A', SessionId)$$
$$IV_{Server2Client} := Hash(K, h, 'B', SessionId)$$
$$EK_{Client2Server} := Hash(K, h, 'C', SessionId)$$
$$EK_{Server2Client} := Hash(K, h, 'D', SessionId)$$
$$IK_{Client2Server} := Hash(K, h, 'E', SessionId)$$
$$IK_{Server2Client} := Hash(K, h, 'F', SessionId)$$

The bits for the key are extracted from the beginning of the output of the cryptographic hash function. If additional bits are required to those produced by the cryptographic hash function, the keys are 'lengthened' using the following method:

Key 'lengthening'

$$K_1 := Hash(K, h, x, SessionId) \text{ with } x = 'A', 'B', ...$$
$$K_2 := Hash(K, h, K_1)$$
$$K_3 := Hash(K, h, K_1, K_2)$$
$$XK := K_1 \, || \, K_2 \, || \, ... \text{ with } XK = IV, EK, IK$$

12.3.3 Client Authentication

The SSH authentication protocol verifies the identity of the client and is run over the the SSH transport protocol. The default supports the following authentication methods:

Supported authentication methods

❏ *Public key:* The client generates a signature that is created with the private key of the user $-K_{User}$ and sends the follow-

Public key

ing message to the server:

$$Client \rightarrow Server: \quad \{SessionId, 50, ID_{User}, Service,$$
$$'public-key', True, PubKeyAlgName,$$
$$+ K_{User}\}_{-K_{User}}$$

Password

❏ *Password:* With this method the password of the client is presented to the server. With this authentication method there has to be an assurance that the SSH connection was negotiated with encryption so that the password cannot be sniffed during transmission.

Host-based public key

❏ *Host-based public key:* Similar to the public key method described above but with the key of the client computer being used.

None

❏ *None:* This method is mainly used to query the schemes supported by the server. If the server does not demand client authentication, it can also respond with a 'success' message.

If verification of the authentication message is successful, the server responds with the message *SshMsgUserauthSuccess*.

12.3.4   Connection Control Within A Session

Connection control within an SSH session is run by the SSH connection protocol that provides the following services:

❏ interactive login sessions;

❏ remote command execution;

❏ encapsulation and routing of TCP/IP connections;

❏ encapsulation and routing of X11 connections.

For each of the services listed above one or more channels, which are all multiplexed into a single encrypted and integrity-protected SSH transport connection, are constructed between client and server. Each of the two peer entities can request that a new channel be opened. Numbers identify the separate channels at sender and receiver and each channel is typed: 'session', 'X11', 'forwarded-TcpIp', etc. Flow control using a sliding window mechanism is performed for each channel and data can only be sent if a window credit is granted by the receiving side.

Opening of a channel

The message *SshMsgChannelOpen* message with the following parameters opens a new channel:

❏ *Channel type* is a character string that specifies one of the channel types listed above.

❏ *Sender channel* is a 32-bit long integer value for the local identification of a channel and is provided by the requesting entity.

❏ *Initial window size* defines the initial sending credit granted by the initiator of the channel of the responding entity.

❏ *Maximum packet size* defines the maximum PDU size that the initiator wants to receive over this channel.

❏ Other parameters are possible and are specific to the requested channel type.

If the receiver of this message does not want to accept the channel request, it responds with the message *SshMsgChannelOpenFailure* that is parameterised as follows:

Rejection of a channel request

❏ *Recipient channel* contains the identification number defined by the sender of the OpenRequest message.

❏ *Reason code* carries one of the values defined in [YKS⁺01b] for signalling the reason.

❏ *Additional textual information* enables the sending of an error message that is readable by the user.

❏ *Language tag* is a character string indicating the language in which the textual information is composed.

If the receiver of an OpenRequest message agrees to opening a new channel, it sends the message *SshMsgChannelOpenConfirmation* with the following parameters:

Acceptance of a channel request

❏ *Recipient channel* contains the identification number allocated by the initiator to the channel.

❏ *Sender channel* indicates an identification number defined by the responder.

❏ *Initial window size* is the initial sending credit granted to the initiator of a channel.

❏ *Maximum packet size* defines the maximum packet size the responder will accept in a channel.

❏ Other parameters can be contained and are specific to the channel type requested.

After a channel has been opened successfully, data transfer is possible, channel-specific requests can be made and the channel can finally be closed. In addition, both peer entities regularly send an appropriate message to renew the sending credit.

Communicating how data should be handled
Before data is sent, channel-specific requests may have to be negotiated in order to agree upon how the data should be handled on the receiving side. This requires the message *SshMsgChannelRequest*. As parameters it contains the identification number of the channel being requested on the receiving side, a RequestType that is formatted as a character string, an indication of whether an explicit response is expected to the request (WantReply) and possibly other parameters that are specific to the requested RequestType. The responding entity can use the two messages *SshMsgChannelSuccess* and *SshMsgChannelFailure* to report on the processing results of the request. The establishment and configuration of specific channel types is explained using examples below.

Establishing an interactive session and starting a command interpreter
The first example illustrates how an interactive session is established with the starting of a command interpreter. Figure 12.6 shows the message exchanges involved. First a 'session' type channel is opened. In the example shown the initiator of the channel defines the local identification number '20' for the channel, grants an initial sending credit of 2048 bytes and specifies a maximum packet size of 512 bytes. The responder replies to this message referring to identification number 20 and defines the value 31 as its own local identification number. It grants an initial sending credit of 1024 bytes and defines 256 bytes as the maximum size for packets sent to it.

The initiator reponds to this by sending a *SshMsgChannelRequest* message typed 'Pty-Req' to request the establishment of a *pseudo terminal*. As it does not yet expect an explicit response, it sets the value of the message field *WantReply* to 'false'. It then immediately requests that an environment variable be set within the session. The variable *home* should then be set to the value '/home/username'. The initiator asks that a command interpreter be started ('Shell', with Unix systems this results in the start of 'Default Shell', e.g. '/bin/sh', entered for the user in '/etc/password') this time requesting an explicit acknowledgement which confirms whether the processing was successful. The responder acknowledges this request with an *SshMsgChannelSuccess* message.

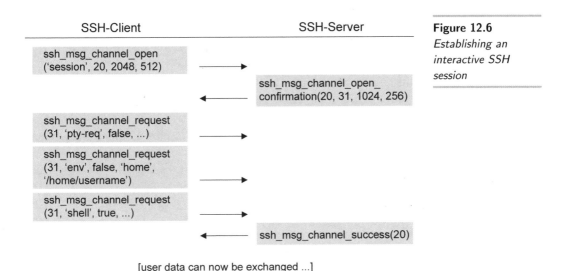

Figure 12.6
Establishing an interactive SSH session

The second example involves tunnelling X11 data streams. The client system, which must have an X11 server running on it, opens a 'session' type channel. It then sends an *SshMsgChannelRequest* message typed *'X11-Req'* to request the tunnelling and routing of X11 data streams. If an X11 application is started on the server later, then its input/output must be encapsulated by the SSH process and sent to the client system. The SSH process on the client system routes this input/output to the local X11 server that has the 'impression' the X11 application is running on the local computer. A separate 'X11' type channel is opened for the data of each X11 application. The server system takes the initiative for this because it is the first system to have knowledge of the application.

Tunnelling of X11 data streams

The final example relates to the tunnelling and routing of TCP/IP traffic, where a distinction is made for the client between incoming and outgoing connections.

Tunnelling and routing of TCP/IP traffic

For outgoing connections the client does not need to make an explicit request for tunnelling in advance and instead can then directly request that the server opens a new 'direct-Tcp' type channel. The channel-type-specific parameters specified by the client are the IP address and the port of the desired target system along with the IP address and the port of the source system.

Outgoing connections

In contrast, with incoming connections the client must send an explicit request for tunnelling to the server so that it has knowledge of this requirement. The client consequently sends an *SshMsgGlobalRequest* message typed 'Tcpip-forward'. In the

Incoming connections

type-specific parameters the client specifies the source addresses from which connections should be accepted (the address '0.0.0.0' indicates that arbitrary systems can be set) and under which port number the server should accept the connections. If a connection setup request for this port arrives at the server later, the server opens a new 'forwarded-Tcpip' type channel. The parameters conveyed to the client are the port number where the connection arrived along with the IP address and the port of the system where the connection setup originated.

12.4   Summary

Within the meaning of the OSI model, the protocols SSL/TLS and SSH, which are usually referred to as *transport layer security protocols*, are actually session layer protocols, as clearly demonstrated by the mechanisms for session control contained in both protocols.

SSL/TLS and SSH are equally suitable for securing Internet communication above the transport layer

SSL/TLS and SSH are equally suitable for securing Internet communication above the transport layer. Both protocols operate on a reliable transport service provided by the TCP protocol. Security protocols for securing connectionless transport protocols such as UDP have not yet been specified and nor implemented to any extent, which means that the security protocols of the lower protocol layers, such as IPSec, L2TP, etc., have to be relied upon for securing connectionless data streams.

Computer-based authentication

Although SSH operates above the transport layer, server authentication is computer-based and is not executed with reference to a specific application process. In this respect SSL/TLS supports multiple certificates and consequently multiple identities within a server system.

The current security protocols of the transport layer implement secure end-to-end communication between application processes. SSH additionally allows application process-related tunnelling and routing of TCP data streams. Furthermore, to a limited degree both protocols can also interoperate with the packet filtering in Internet firewalls (see also Chapter 13) and in many cases present a very good option for securing Internet communication.

Limited firewall compatibility

However, as these protocols in principle cannot secure the protocol headers of the lower communication layers, they offer no protection against attacks on actual network infrastructure and therefore cannot replace but only enhance the security protocols of the lower communication layers.

12.5   Supplemental Reading

[BKS98]   BLEICHENBACHER, D.; KALISKI, B.; STADDON, J.:
              *Recent Results on PKCS #1: RSA Encryption Stan-
              dard.* 1998. – RSA Laboratories' Bulletin 7

[CFPR96]   COPPERSMITH, D.; FRANKLIN, M. K.; PATARIN, J.;
              REITER, M. K.: Low Exponent RSA with Related Mes-
              sages. In: MAURER, U. (Hrsg.): *In Advances in Cryp-
              tology – Eurocrypt'96 Proceedings*, Springer, 1996. –
              Vol. 1070 of Lectures Notes in Computer Science

[FKK96]   FREIER, A. O.; KARLTON, P.; KOCHER, P. C.: *The SSL
              Protocol Version 3.0.* 1996. – Netscape Communica-
              tions Corporation

[DA99]   DIERKS, T.; ALLEN, C.: *The TLS Protocol Version
              1.0.* January 1999. – RFC 2246, IETF, ftp://ftp.
              internic.net/rfc/rfc2246.txt

[YKS+01a]   YLONEN, T.; KIVINEN, T.; SAARINEN, M.; RINNE, T.;
              LEHTINEN, S.: *SSH Authentication Protocol.* 2001.
              – Internet draft (work in progress), draft-ietf-secsh-
              userauth-11.txt

[YKS+01b]   YLONEN, T.; KIVINEN, T.; SAARINEN, M.; RINNE,
              T.; LEHTINEN, S.: *SSH Connection Protocol.* 2001.
              – Internet draft (work in progress), draft-ietf-secsh-
              connect-11.txt

[YKS+01c]   YLONEN, T.; KIVINEN, T.; SAARINEN, M.; RINNE, T.;
              LEHTINEN, S.: *SSH Protocol Architecture.* 2001.
              – Internet draft (work in progress), draft-ietf-secsh-
              architecture-09.txt

[YKS+01d]   YLONEN, T.; KIVINEN, T.; SAARINEN, M.; RINNE, T.;
              LEHTINEN, S.: *SSH Transport Layer Protocol.* 2001.
              – Internet draft (work in progress), draft-ietf-secsh-
              transport-09.txt

12.6   Questions

1.   Which security services are provided by the SSL protocol?

2. Why is it with SSL that it is often only the server that authenticates itself to the client?

3. Which concepts of the session layer based on the OSI model are implemented by SSL?

4. The default in the SSL record protocol does not provide for data compression. Why is this an advantage?

5. Which other security property is protected in the SSH transport protocol as a result of the MAC being computed over the plaintext before the plaintext is encrypted (instead of the equally conceivable sequence in which encryption is performed before the MAC is computed)?

6. In an exchange based on the handshake protocol, what convinces a client of the authenticity of the server when key negotiation uses Diffie–Hellman?

7. Which cryptographic property is referred to 'plaintext aware'?

8. To which layers of the OSI model can the SSH protocol be related? Give the appropriate protocol functions for each instance.

9. What advantage is derived from the fact that specific Diffie–Hellman parameters are specified for SSH? Are there also potential disadvantages?

10. Why does SSH provide for password-based authentication of the client?

11. What is the purpose of the sending credits granted in SSH? Could this function not be handled by the underlying transport protocol?

12. Compare SSH with SSL (listing similarities as well as differences).

13. Compare the methods for 'extending' the length of session keys of SSH and the Internet Key Exchange (Chapter 11) with respect to the entropy of derived session keys.

13 Internet Firewalls

The preceding chapters in this part of the book mainly deal with protecting protocol data units from attack during transmission. This chapter expands on that discussion and is devoted to the task of protecting certain parts of a network from the intrusion of 'undesirable' data units.

In building construction, the word *firewall* has become the established term for describing the task of protecting certain parts of a building from the spread of fire. Based on this term, the components used for protecting specific subnetworks from potential attack from other parts of the Internet are referred to as *Internet firewalls*. This term may not be totally accurate from a technical standpoint since Internet firewalls, unlike normal firewalls, should not be completely impenetrable (disconnecting the physical network connection would suffice in this case). However, it does convey an intuitive understanding of the necessary task.

Similar term as for firewalls in building construction

13.1 Tasks and Basic Principles of Firewalls

Due to the fact that an Internet firewall should not be completely impenetrable, it can easily be compared to the drawbridge of a medieval castle [ZCC00]. A drawbridge typically had the following function:

Tasks

❏ It forced people to enter a castle at one specific location that could be carefully controlled.

❏ It prevented attackers from getting close to other defence components of the castle.

❏ It forced people to exit the castle at one specific location.

For obvious reasons, an Internet firewall is normally installed at the network gateway between a protected trusted subnetwork and an untrustworthy network, which allows it to monitor incoming and outgoing data traffic. An example of this is the point where a corporate local area network is connected to the global Internet (compare Figure 13.1). All systems of a protected subnetwork are normally protected in the same way, i.e. from the same undesirable data packets with access control implemented by an Internet firewall at the subnetwork level.

Internet firewalls provide access control at the subnetwork level

Figure 13.1
Firewall placement between a protected network and the Internet

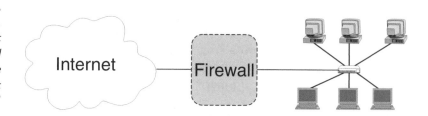

Functions of Internet firewalls

An Internet firewall therefore provides a central point where access control decisions are made and implemented [Sem96]. It is also an effective means of monitoring the Internet activities of subnetworks and preventing certain security problems from proliferating from one network area to another.

Limit to protection provided by Internet firewalls

Firewalls can – if at all – only provide limited protection against malicious insiders who are determined to bypass the security strategy of a subnetwork. In particular, a firewall cannot influence any data traffic that is not passing through it. This self-evident fact is often overlooked. An example for this is a modem dial-up access that is set up with a system placed behind the corporate firewall. A firewall furthermore cannot protect against totally new attack patterns and also only offers limited protection against the intrusion of software viruses.

A differentiation is made between two diametrically opposed approaches to the basic security strategy that can be implemented by an Internet firewall [WC95]:

Default-Deny Strategy

❏ The *Default-Deny Strategy* is based on the principle that 'Anything that is not explicitly permitted is denied'. First an analysis is undertaken to determine which network services users of the protected subnetwork require and how these services can be made available in a secure way. Only those services for which a legitimate need exists and that can be supplied in a secure manner are provided. All other services are blocked by technical means. An example of this approach is permitting the transmission of HTTP traffic to any arbitrary systems and SMTP traffic to an e-mail gateway. All other data traffic would be blocked by the firewall. The advantage of this strategy from the perspective of security is that it also provides limited protection from unknown threats.

❏ The *Default-Permit Strategy* follows an opposite approach, which is 'Anything that is not explicitly forbidden is permitted'. With this approach services categorised as high risk or contravening local security strategies are explicitly blocked by appropriate measures in the firewall. All other services are considered permissible and not blocked by the firewall. For example, such a strategy could mean that traffic from the *Network File System (NFS)* and *X11* protocols are generally blocked and that Telnet connections are only permitted to a specific host. This is the approach most users prefer because it usually gives them more flexibility. However, this strategy can, in principle, only offer protection from known threats.

Default-Permit Strategy

As the description of these two approaches indicates, each Internet service blocked or permitted by a firewall always has to be examined specifically. The following section provides a brief overview of popular Internet services and their transport within the framework of the TCP/IP protocol family.

13.2 Firewall-Relevant Internet Services and Protocols

In practice, the following Internet services are of particular importance in the context of Internet firewalls:

Internet services

❏ *Electronic mail (e-mail)*, normally implemented with the *Simple Mail Transfer Protocol (SMTP)*.

❏ *File exchange* on the basis of the *File Transfer Protocol (FTP)*, *Network File System (NFS)* and *NetBIOS* protocols.

❏ *Remote terminal access* and *remote command execution* through *Telnet, RLogin* or *Secure Shell (SSH)* protocols.

❏ *Usenet News*, which is distributed through the *Network News Transfer Protocol (NNTP)*.

❏ *World Wide Web* on the basis of the *Hypertext Transfer Protocol (HTTP)*.

❏ *Information about people* that can be called up using the *Finger* protocol.

❏ *Real-time conferencing systems*, such as *CUseeMe, Microsoft Netmeeting, Netscape Conference* and the multicast-based *MBone Tools*.

❏ *Name services*, which are provided in the Internet through the *Domain Name System (DNS)*.

❏ *Network management* based on the *Simple Network Management Protocol (SMTP)*.

❏ *Clock synchronisation* using the *Network Time Protocol (NTP)*.

❏ *Graphical window systems* with *X11* being the most popular protocol in use.

❏ *Printing services*, based, for example, on the *Line Printing Protocol (LPR)*.

These services are normally realised through client/server programs and the application protocols used between these programs. The data units of the application protocols are usually transported either in the segments of a TCP connection or in the datagrammes of the UDP protocol. TCP segments and UDP datagrammes are transported in IP packets that in turn are sent in protocol data units of the Layer 2 protocols used on the different communication links between source and destination computer (for example, Ethernet, FDDI, ATM).

Addressing of application processes

The two tuples *(source-IP address, source port)* and *(destination-IP address, destination port)* are used to address individual application processes. A *port* represents a number, comprising two octets, which explicitly identifies the Service Access Point (SAP) of an application process within an end system.

Firewall-relevant protocol fields

Figure 13.2 shows the frame format of an IP packet that is transporting a TCP segment. The following protocol fields are of particular importance for Internet firewalls:

❏ Fields of the access protocol ('Layer 2 protocol', not shown): The fields of particular interest are those that identify the contained Layer 3 protocol (e.g. IP, Appletalk, IPX) and the source addresses of the access protocol (e.g. Ethernet-MAC address).

❏ Fields of the IP protocol: Of special interest here are the source and destination addresses, the field Flags, which also contains one bit to display IP fragments, the protocol identification of the user data contained (e.g. TCP, UDP) and possible options such as the stipulation of the transmission path by the source. However, IP options are seldom used for anything other than attacks.

❏ Fields of the TCP protocol: The fields of particular interest are the source and destination ports that have a limited influence on determining the sending and receiving application, since many popular Internet services use well-defined port numbers. In addition, individual bits of the control field have an effect on connection control. An ACK bit is set in all PDUs except in the first connection establishment PDU, which means that an ACK bit that has not been set identifies a connect PDU. A SYN bit is set in all PDUs except in the first two PDUs of a connection. A packet with a set ACK bit and a SYN bit that is not set therefore identifies the PDU accepting the connection. Lastly, an RST bit permits a connection to be terminated immediately without displaying a further error message, and can therefore be used to terminate a connection without providing additional information.

❏ Fields of the application protocol: In some instances it may also be necessary or meaningful for the Internet firewall to check the protocol fields of an application protocol. However, these vary from application to application and therefore will not be discussed here.

13.3 Terminology and Building Blocks

Internet firewalls are usually made up of certain building blocks that fulfil specific tasks and possess certain characteristics. This section briefly introduces the relevant terminology.

An *Internet firewall* (hereafter also simply referred to as firewall for short) is a component, or a group of components, that restricts access between a protected network and the Internet or another untrustworthy network. In this context access refers to the exchange of data packets. Even if the term may appear to be an unusual choice at first glance, it should convey that, from the view of security, access control decisions are made and implemented on the basis of a firewall.

Internet firewall

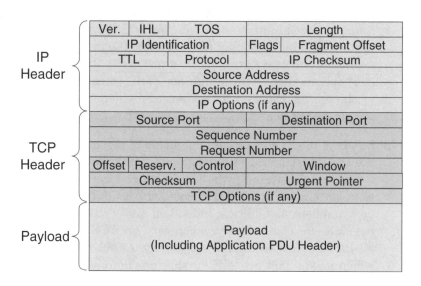

Figure 13.2
Frame format of an IP packet with a TCP segment

Packet filtering

Packet filtering specifies the actions taken by a specific system when selectively controlling the flow of data packets from and to a specific network. Packet filtering is an important basic technique for implementing access control at the subnetwork level in packet-oriented networks such as the Internet. The packet filtering process is also referred to as *screening*.

Bastion host

A *bastion host* is a computer that has special security requirements due to the network configuration that makes it more vulnerable to certain attacks than the other computers in a subnetwork. A bastion host within a firewall often represents the main point of contact between application processes within a protected network and the processes of external hosts.

Dual-homed host

A computer is identified as a *dual-homed host* if it has at least two network interfaces and is therefore 'at home' in two subnetworks.

Proxy

A *proxy* is a program that deputises for the client processes of a protected subnetwork and communicates with external server programs. Proxies relay approved requests from internal clients to the actual servers and also forward server responses to the internal clients.

Perimeter network, demilitarized zone (DMZ)

The term *perimeter network* identifies a subnetwork that is inserted between an external and an internal network in order to create an additional security zone. Perimeter networks are often referred to as *demilitarized zones, DMZ)*.

Network Address Translation (NAT) is a technique in which an intermediate system deliberately modifies the address fields in a data packet in order to implement an address conversion between internal and external addresses. For example, with this technique a large number of systems can be connected to the Internet on the basis of a small number of externally valid addresses (i.e. in the Internet). Although NAT is not actually a security technique, it offers the advantage from a security perspective that the internal addresses will not be known to external attackers. This makes it more difficult for deliberate attacks to be executed to individual internal systems from the outside. With NAT the 'initiative' for specific data streams must basically originate from internal systems, and the application processes of internal systems are not visible from the outside.

Network Address Translation (NAT)

13.4 Firewall Architectures

As mentioned above, Internet firewalls can be built from one component or a group of components. As a result, a number of the firewall architectures have established themselves over the years. The structures of some of the popular ones will be explained in this section.

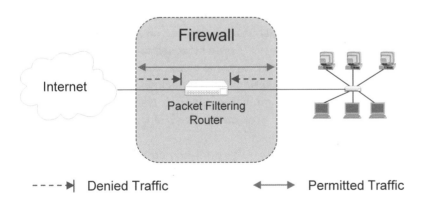

Figure 13.3
Architecture of a packet filter firewall

The simplest of these architectures consists solely of one packet filter and is illustrated in Figure 13.3. This architecture can be built either with a commercially available workstation computer using two network interfaces or with a dedicated router that also incorporates basic packet filter functions. The protection function of this architecture is based solely on filtering undesirable data packets between a protected network and the Internet (or other

Protection using simple packet filtering

untrustworthy subnetwork). In Figure 13.3 (and in other figures in the rest of this section) a straight arrow shows the 'permitted' data traffic and a dotted line the 'blocked traffic'.

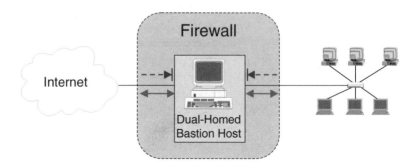

Figure 13.4 illustrates the minimally more complex *dual-homed host architecture*. Although there may be no difference between the hardware configuration of this architecture and a packet filter (if the packet filter is implemented on a workstation computer), the software on the dual-homed host provides additional security functions because its main task is not limited to pure packet filtering.

A dual-homed host provides proxy services

A dual-homed host provides proxy services to internal and external clients, so for security reasons IP packets do not need to be routed directly into and out of the protected subnetwork. The protected network in this case is only attached to the Internet through monitored proxy services. In some cases, the dual-homed host can also handle routing functions for IP packets. However, appropriate packet filter functionality must then also be implemented on the host to prevent the protective effects of the firewall from being lost.

A dual-homed host should be secured as a bastion host

With this architecture the dual-homed host is the central attack point for potential attackers. Therefore, it should be regarded as a bastion host and protected accordingly (compare also Section 13.6).

The drawback of this architecture is that, depending on the size of the protected subnetwork and the bandwidth of the Internet connection, a dual-homed host can develop into a performance bottleneck.

Screened-host architecture

Figure 13.5 shows a *screened-host architecture*, which spreads the functionality of the packet filter and the proxy server over two components. The packet filter allows permitted traffic to pass be-

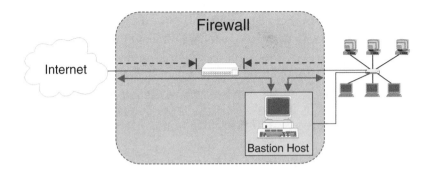

Figure 13.5
Screened-host architecture

tween the Internet and the internal bastion host and blocks all direct traffic between other internal computers and the Internet. In some cases, certain traffic streams can also be permitted directly between the internal computers and the Internet; however, this can introduce potential vulnerabilities. The bastion host in turn provides proxy services for communication between internal computers and the Internet.

Two main advantages can be gained from separating packet filter functionality and proxy services into two consecutively switched devices:

Advantages of separating packet filtering and proxy services

❑ In terms of security, a bastion host will be better protected from attacks on potentially available services that should not be accessible from outside. Furthermore, compared with proxy services, relatively simple software can be used to implement packet filter functionality, which means that a relatively compact software installation is all that is required on the respective system. This is particularly an advantage because complexity is the main obstacle in the construction of secure systems.

❑ Another benefit of this architecture is that it can also provide a performance advantage. When certain data streams (e.g. Internet telephony) are routed directly into an internal network, they do not have to 'cross through' the same computer that also provides the proxy services.

In terms of the physical structure of the networking, it should be noted that a bastion host could sometimes also be connected to the same internal subnetwork for which it provides security services. In this case, its relationship to the firewall tends rather to be of a logical nature since it is only conditional on the packet filter rules.

Physical networking structure

It is particularly important to note that a bastion host essentially has special protection needs and should therefore be more difficult to access physically than the systems protected by it (also see Section 13.6). However, this kind of networking presents a security problem even if the bastion host is a physically separate installation: depending on the networking technique used (e.g. Fast-Ethernet-Hub), attackers who are able to compromise a bastion host can then directly sniff the traffic in the internal network.

Sniffing threat when bastion host compromised!

Figure 13.6
Screened-subnetwork architecture

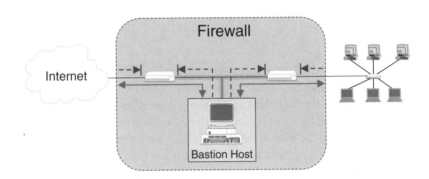

The *screened-subnetwork architecture* illustrated in Figure 13.6 counters the particular threat of a potentially compromised bastion host by setting up a *demilitarised zone*, also referred to as a *perimeter network*. This zone is created using two packet filters between which one or more bastion hosts is embedded for the purpose of providing proxy services. This zone is also available for the connection of the servers that provide services to external computers, such as WWW and FTP. The main task of the second packet filter is to protect the internal network from the threat of a possibly compromised bastion host.

Perimeter network

The *split-screened subnetwork architecture* shown in Figure 13.7 is an even more sophisticated architecture in terms of security. In this architecture, a dual-homed bastion host divides the demilitarised zone into two separate subnetworks. Using proxy services, the bastion host is better able than a simple packet filter to exercise finely granulated control over the data streams. Furthermore, an outer packet filter protects the bastion host from external attackers and an inner packet filter protects the internal subnetwork from the bastion host if it is compromised. The architecture thus provides an 'in-depth defence', comparable to that of a knight's castle where attackers normally have to surmount several protective walls before they can penetrate the interior.

Split-screened subnetwork architecture

'In-depth defence'

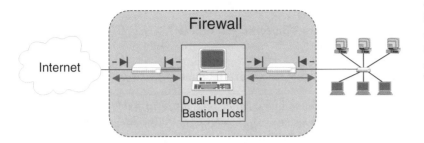

Figure 13.7
*Split-screened
subnetwork
architecture*

13.5 Packet Filtering

As our discussion on the different firewall architectures has shown, the packet filtering function plays a central role in securing individual subnetworks. This section takes a detailed look at the role of packet filtering.

The first question in this context is, which protective functions can IP packet filtering provide for access control?

Theoretically, protective measures of any complexity are feasible because all data exchanged between two or more entities in the Internet is ultimately transported in IP packets. In practice, however, the two basic considerations listed below have proven useful in the decision criteria for designing protective measures:

Proxy service or packet filtering?

1. Selection operations, which require detailed knowledge about the functioning of higher level protocols or a comprehensive evaluation of the protocol data units transmitted in previous IP packets, are easier to implement as specific proxy services.

2. Simple selection operations that have to be executed quickly and always evaluate individual IP packets can be implemented more efficiently through the use of packet filters.

This kind of *basic packet filtering* uses the following information to make its selection decision:

Basic packet filtering

❑ source IP address;

❑ destination IP address;

❑ transport protocol;

❑ source address port;

❏ destination address port;

❏ in some cases, specific protocol fields (e.g. TCPs ACK-bit and SYN-bit);

❏ network interface where IP packet was received.

Stateful packet filtering

In addition to these basic functions, *stateful packet filtering* is also sometimes used in selection decisions. This variant, also referred to as *dynamic packet filtering*, ensures that incoming UDP packets are only allowed to pass if they are a response to previously observed outgoing packets.

Basic protocol check

Lastly, packet filters can also handle basic *protocol checks*. An example is the checking of IP packets that are sent to a DNS port for correct formatting in accordance with the DNS protocol. Another related point is that packets containing an HTTP-PDU in their data part should definitely not be forwarded to certain IP addresses.

Increasing complexity means additional resources!

The design of packet filter functions should take into account that the more complex they become, the more resources they require for execution. Therefore, it is always important to consider whether a two-step approach using a simple packet filter and application-specific proxy services would produce a more efficient mechanism for the checking process.

Packet filter actions

Based on the filtering rules configured in it, a packet filter decides how to proceed with each IP packet it looks at. This involves the following actions:

❏ forwarding the packet;

❏ dropping the packet;

❏ in some cases, logging either an entire packet or parts of it;

❏ possibly sending an error message to the sender, taking into account that error messages could be helpful to any attackers who are preparing further attacks.

Specification of basic filter rules

Because packet filters normally attempt to separate a protected subnetwork from a less trustworthy network, an implicit understanding of the traffic direction is required for the specification of filter rules. Therefore, IP packets that are received from a network interface located outside the protected network are referred to as *inbound traffic*. Accordingly, packets that are received from the network interface of a protected subnetwork are considered to be *outbound traffic.* When packet filter rules are specified, the information identifying which network interface received the IP packet is usually based on one of the three alternatives *'inbound', 'outbound'* or *'either'*.

Placeholders (also called 'wildcards') can be used to specify source and destination IP addresses: for example, '125.26.*.*' refers to all IP addresses that begin with the prefix '125.26.'.

Use of 'wildcards'

The IP addresses in the following examples are often simply indicated as *'internal'* or *'external'* to keep the discussion neutral in terms of any particular network topology. Often, when the source and destination ports are given, entire ranges are specified, e.g. '>1023' for all port numbers higher than 1023.

One basic assumption is that packet filter rules are evaluated in the sequence in which they are specified and that the first rule applicable to a packet determines the action to be taken. Over the years it has been proved that this method is the easiest one for system administrators to understand and the best one for keeping errors in rule specification to a minimum.

Sequence of evaluating packet filter rules

The set of packet filter rules developed below illustrate the concepts described so far and are aimed at ensuring that only e-mail traffic is allowed between a protected subnetwork and the Internet(example according to [ZCC00]).

Example of a packet filter specification

Internet e-mail is being exchanged between two SMTP servers in TCP connections. The initiative for the exchange originates from the server that is delivering the e-mails. This server establishes a TCP connection to destination port 25 of the receiving SMTP server for this purpose, selecting a number higher than 1023 as the source port. Since incoming as well as outgoing e-mails are allowed, packet filter rules have to be designed for both directions.

Table 13.1 shows the first approach that can be taken. The two rules A and B are used to transmit incoming e-mails and rules C and D fulfil the same task for outgoing e-mails. The last one, Rule E, is aimed at denying all other data traffic ('Default-Deny Strategy').

First approach

Rule A allows inbound IP packets that are being sent to destination port 25 and are transporting TCP segments as payload. For the TCP connection to be established and maintained between the two SMTP servers the corresponding response packets of the server receiving the emails must be allowed to pass outside. Rule B therefore allows outbound IP packets to port numbers higher than 1023.

Table 13.1

*Example of a packet
filter specification*

Rule	Direction	Src. Addr.	Dest. Addr.	Protocol	Src. Port	Dest. Port	ACK	Action
A	Inbound	External	Internal	TCP		25		Permit
B	Outbound	Internal	External	TCP		>1023		Permit
C	Outbound	Internal	External	TCP		25		Permit
D	Inbound	External	Internal	TCP		>1023		Permit
E	Either	Any	Any	Any		Any		Deny

*Protection function of
the first approach*

The following examples help to explain the protection function of these rules:

❏ Assume that an IP packet with a forged internal source address is being sent to the protected subnetwork from the outside: Since inbound packets are only forwarded if they have external source addresses and internal destination addresses (rules A and D), this sort of attack will be successfully defeated. The same applies to outgoing packets from an internal attacker making use of external source addresses (rules B and C).

❏ This filter specification is also effective at blocking incoming Telnet traffic, because Telnet servers normally wait at port 23 for incoming connections and incoming traffic is only allowed by the packet filter if it is being sent either to port 25 or to a port higher than 1023 (rules A and D). On the basis of rules B and C, the same applies to outgoing Telnet traffic.

*Drawback of first
approach*

However, this ruleset is not totally effective. For example, it does not block either incoming or outgoing X11-Protocol traffic. An X11 server usually waits at port 6000 for incoming connections and the corresponding client programs use destination ports from the set of numbers above 1023. Because of Rule B, inbound X11-PDUs can therefore pass the packet filter and, because of Rule D, outbound X11-PDUs are not blocked either. This provides potential attackers with an attractive opportunity because the X11-Protocol leaves the door open to various different attacks, e.g. reading and manipulation of screen information and of keyboard input.

*Second approach:
Including a source
port*

This problem can be dealt with through the inclusion of the source port in the filter specification. Table 13.2 shows the resulting filter rules. With this ruleset outbound packets are only permitted to ports higher than 1023 if they originate from source

Rule	Direction	Src. Addr.	Dest. Addr.	Protocol	Src. Port	Dest. Port	ACK	Action
A	Inbound	External	Internal	TCP	>1023	25		Permit
B	Outbound	Internal	External	TCP	25	>1023		Permit
C	Outbound	Internal	External	TCP	>1023	25		Permit
D	Inbound	External	Internal	TCP	25	>1023		Permit
E	Either	Any	Any	Any	Any	Any		Deny

Table 13.2
Inclusion of source port in a packet filter ruleset

port 25 (Rule B). Therefore, PDUs from internal X11-client or X11-server programs are blocked. Similarly, Rule D blocks incoming traffic to X11-client or X11-server programs.

In practice, however, one cannot assume that an attacker will not use Port 25 as the source port for an attacking X11-client. In this instance the packet filter would allow the traffic to pass.

Drawback of second approach

Rule	Direction	Src. Addr.	Dest. Addr.	Protocol	Src. Port	Dest. Port	ACK	Action
A	Inbound	External	Internal	TCP	>1023	25	Any	Permit
B	Outbound	Internal	External	TCP	25	>1023	Yes	Permit
C	Outbound	Internal	External	TCP	>1023	25	Any	Permit
D	Inbound	External	Internal	TCP	25	>1023	Yes	Permit
E	Either	Any	Any	Any	Any	Any	Any	Deny

Table 13.3
Inclusion of an ACK bit in the packet filter ruleset

Table 13.3 therefore shows an improved version of the filtering specification that also incorporates the ACK bit of the TCP protocol header and includes rules B and D. Since the ACK bit is assumed to be set in Rule B, this rule can no longer be used to open an outgoing connection (the TCP connect-request is identified by an ACK bit not set). Similarly, the stipulation of a set ACK bit in Rule D prevents the opening of incoming TCP connections sent to ports with a number higher than 1023.

Third approach: Including a TCP-ACK bit

As a basic guideline it should be noted that each packet filter rule designed to allow TCP-PDUs for outgoing connections (or the reverse) should request a set ACK bit.

If the firewall contains a bastion host, then the SMTP server should preferably be run on this system (or the SMTP host be regarded as a bastion host). Incorporating the IP address of the bastion host into the packet filter ruleset can increase the security of

Fourth approach: Including address of SMTP server

the protected subnetwork, because in this instance attacks on the SMTP server program will only affect the bastion host. Table 13.4 shows the resulting packet filter ruleset.

Table 13.4

Inclusion of bastion host in packet filter ruleset

Rule	Direction	Src. Addr.	Dest. Addr.	Protocol	Src. Port	Dest. Port	ACK	Action
A	Inbound	External	Bastion	TCP	>1023	25	Any	Permit
B	Outbound	Bastion	External	TCP	25	>1023	Yes	Permit
C	Outbound	Bastion	External	TCP	>1023	25	Any	Permit
D	Inbound	External	Bastion	TCP	25	>1023	Yes	Permit
E	Either	Any	Any	Any	Any	Any	Any	Deny

With a screened subnetwork firewall, two packet filters have to be equipped with the appropriate rulesets: one for the traffic between the Internet and the bastion host and one for the traffic between the bastion host and internal subnetwork.

13.6   Bastion Hosts and Proxy Servers

Point of contact between internal and external systems

As explained in Section 13.3, the bastion host is the main point of contact between application processes within a protected network and the processes of external systems. Depending on the firewall architecture, it is responsible for the task of packet filtering and/or provides dedicated proxy services for specific applications. As the principal point of contact for the protected network to external systems, a bastion host has a higher exposure to threats than the systems to which it provides security functions.

Even though a dedicated packet filter can be used to reduce these risks, special measures should also be taken to secure the proxy server itself. This section begins by presenting some general observations on securing bastion hosts and concludes by briefly reviewing how proxy services are implemented.

Configuration of a bastion host

The principles applied to building a 'secure' bastion host are ultimately merely extensions to the strategies used for securing any key system in a network. The basic rule is that the system configuration of a bastion host should be kept *as simple as possible*. If there are any services that should not be offered on the bastion host, they should not even be installed on it.

Potential compromising of bastion host

One should also basically anticipate the fact that a bastion host might be compromised. Internal systems should therefore not place more trust in the bastion host than is absolutely necessary.

For example, no file systems should be exported to the bastion host, no Login should be enabled from the bastion host to internal systems, etc. If possible, the bastion host should be integrated into the network infrastructure in such a way that it is unable to eavesdrop on data traffic in the internal subnetwork. (This can be done through the use of another packet filter to separate the network or through a network interface that does not support 'promiscuous mode', which is necessary for eavesdropping).

In addition, extensive event logging should take place on the bastion host to ensure early detection of any attacks or successful compromising of the host. Note that a successful attacker should not be able to tamper with event logging at a later time. One possibility is to transfer the event entries over the serial interface to a separate log-computer that does not have a network interface itself.

Event logging for attack detection

As long as no important reasons exist to the contrary, the bastion host should preferably be an 'unattractive' target for potential attackers. For example, slow computers are less attractive targets and also of less use to an attacker than high-speed systems. At the same time one has to bear in mind that a bastion host should not turn into a performance bottleneck for the internal network and sometimes also has to offer resource-intensive services (e.g. Web proxy services). None the less, the potential usefulness of a bastion host to an attacker should be minimal. Therefore, the only software tools that should be installed on it are those that are essential to its operation. Above all, no user accounts should be available on a bastion host. This will prevent anyone from spying on data or executing password attacks.

Unattractive attack target

Lastly, a bastion host should definitely be installed in a secure location where general physical access is not allowed to ensure that the risk of manipulation is kept to a minimum. Its system configuration should be backed up at regular intervals. A routine reinstallation of the system from a security copy can also be considered for the purpose of correcting any possible manipulation that has occurred. However, it is important to be aware that any existing weakness in a system configuration that has already been exploited by an attacker can be used again for gaining access to the system. A reinstallation of the system therefore does not reduce the risk of it being compromised.

Installation in a secure location

A bastion host equipped with proxy services can give users of a protected subnetwork the illusion that all systems in this network

Proxy services

are able to access Internet services although in reality it is only the bastion host that has access.

There are two main different types of proxy services:

Application level proxy

❏ If a proxy server analyses the commands of the application protocol and interprets its semantics, it is referred to as an *application level proxy*.

Circuit level proxy

❏ If, on the other hand, the proxy server is restricted to forwarding application PDUs between client and server, it is called a *circuit level proxy*.

Possible 'candidates' for proxy services include FTP, Telnet, DNS, SMTP and HTTP. The following situation usually exists when a proxy server is available: The user of the proxy service thinks that his or her system is exchanging data directly with the actual server, and the server is under the illusion that it is exchanging data with the bastion host.

Different variants of proxy services

If a proxy service is used, applications-specific data streams have to be rerouted to the proxy server. In each case the proxy server must be informed of which application server it is to use for the connection. The following four implementation possibilities exist:

User procedures

❏ *Proxy-aware user procedures:* An example of this variant is the use of Telnet proxies. Here the user first registers on the bastion host and then from there establishes a Telnet session to the actual desired server.

Client software

❏ *Proxy-aware client software:* In this case the client software is responsible for transferring the corresponding data to the proxy server. This approach is used, for example, with proxy-enabled Web browsers where client software only has to be notified once of the name or address of the Web proxy.

Operating system

❏ *Proxy-aware operating system:* With this variant the operating system (using the appropriate configuration data) is responsible for the rerouting and transfers the required parameters to the proxy server.

Routers

❏ *Proxy-aware router:* This variant transfers the task of activating the proxy to an intermediate system, which reroutes connections initiated by internal client computers to a proxy computer, also sending the necessary address data to the computer.

13.7   Summary

Internet firewalls provide access control at the subnetwork level. For this purpose, Internet firewalls are normally installed between a protected subnetwork and a less trustworthy network, such as the public Internet.

Access control at the subnetwork level

At the same time it should be noted that an Internet firewall is only capable of monitoring those data streams that pass through it and that it basically cannot provide any protection from internal attackers.

Depending on the level of security being sought and the resources available, an Internet firewall can be constructed using either a single or a group of components. Over time a number of architectures have evolved as a result, each offering specific security features. In practice, the *packet filter*, the *dual-homed-host*, the *screened-host*, the *screened-subnetwork* and the *split-screened-subnetwork* architectures are the most popular.

Firewall architectures

The central elements used by all these architectures are *packet filters*, which make access decisions for each IP packet routed through them, and *proxy servers*, which enable an applications-specific monitoring of data streams. The computing systems with specific functions installed in a firewall are called *bastion hosts*. Because they have a high level of exposure to attackers, bastion hosts must be installed with particular care and constantly monitored.

Central elements are packet filters and proxy servers

Bastion hosts

13.8   Supplemental Reading

[Sem96]   SEMERIA, C.: *Internet Firewalls and Security*. 1996. – 3Com Technical Paper

[WC95]   WACK, J. P.; CARNAHAN, L. J.: *Keeping Your Site Comfortably Secure: An Introduction to Internet Firewalls*. 1995. – NIST Special Publication 800-10

[ZCC00]   ZWICKY, E.; COOPER, S.; CHAPMAN, B.: *Building Internet Firewalls*. Second Edition, O'Reilly, 2000
A recognised standard work in this field that not only offers in-depth explanations of the principle functions of firewall components but also provides detailed practical information on packet filtering and securing the most popular Internet services.

13.9   Questions

1. What basic information is normally necessary for access control decisions that identify whether an IP packet should be routed or not?

2. What further tests can be carried out and which type of components should be used for this purpose?

3. Does a bastion host basically have to be a dual-homed host? Is the opposite conclusion correct?

4. Why should information identifying which network interface card received a packet be included in the generation of packet filter rules?

5. Use of packet filtering for access control:

   ❏ What is the main underlying assumption in the use of packet filtering for access control?

   ❏ Can this lead to potential security risks?

   ❏ If yes, how can one deal with these risks?

   ❏ Can you give some reasons why the use of packet filters still makes sense?

6. Why should packet filter rules basically be applied in the sequence of their specification?

7. Research (e.g. using the appropriate Internet RFC) the information required for packet filtering of the Telnet protocol and list the packet filter rules for blocking Telnet data streams.

8. Which restriction can you implement in packet filtering through an appropriate evaluation of the ACK bit of TCP?

9. Why do you basically have to allow two traffic directions with TCP-based services?

10. How can you protect your network so that a possibly compromised bastion host is not misused for eavesdropping on your internal traffic?

Part III

Secure Wireless and Mobile Communications

14 Security Aspects of Mobile Communication

Part III of this book is devoted to aspects of security specific to mobile communication. This chapter first examines some general aspects and the subsequent chapters discuss specific system examples. Chapter 15 describes security mechanisms in wireless local area networks based on the IEEE 802.11 standard, focusing particularly on the security shortcomings of the standard and potential alternatives. Chapter 16 deals with the security of radio-based wide-area networks, with specific emphasis on the security mechanisms of GSM and UMTS networks that use similar principles. Chapter 17, the final chapter of the book, is devoted to mobile Internet communication. It discusses the central approaches of Mobile IP and AAA infrastructure (authentication, authorization and accounting), again with particular attention to security.

14.1 Threats in Mobile Communication Networks

Any comprehensive study of the emerging engineering discipline network security will raise the issue of the extent to which the circumstances and technical characteristics of mobile communication introduce new security aspects and solutions for mobile communication networks. The first thing to note is that mobile communication is naturally a target for all the threats that occur in fixed network communication, i.e., masqueraded identities, authorisation violations, eavesdropping, data loss, modified and falsified data units, repudiation of communication processes and sabotage. Consequently, similar measures as in fixed networks should be taken in mobile communication networks.

Do mobile communication networks introduce new security aspects?

However, some aspects do not occur in the same form in fixed networks and are caused by the mobility of users and their devices, as well as by the existence of wireless communication segments.

Specific aspects

Increase of risk potential

❏ Some of the threats that already exist in fixed networks form a greater risk potential for mobile networks: for example, wireless transmission links are easier to eavesdrop than are wireline transmission media due to the ease of gaining direct physical access. Likewise, the lack of a physical connection makes it easy for unauthorised entities to use the services of a wireless network if adequate security mechanisms are not in place.

New difficulties arise in providing security services

❏ Some new difficulties arise in providing security services: for example, the authenticity of a mobile device has to be verified again by the respective network access point each time it changes network access point, i.e., performs a *handover*. Key management is also more complicated because the respective peer entities generally cannot be determined in advance as they depend on the movements of the user.

Movement profiles

❏ Ultimately a completely new threat arises: the current location of a device and, consequently, also of its user provides much more interesting information than in fixed networks and therefore should be protected against eavesdropping.

The first two aspects are handled in detail in the discussion covering specific system examples in the following chapters. In terms of the third aspect, protecting the identity of a user's current location, existing architectures show serious shortcoming in addressing this requirement. This subject is still mostly being dealt with in research and adequate solutions are still being sought for use in mobile communication architectures. The following section, therefore, treats this aspect in general terms independently of any concrete system examples.

14.2   Protecting Location Confidentiality

As the following chapters show, the mobile communication networks that operate today do not incorporate effective measures for adequately protecting information about the current location of mobile devices *(location privacy)*. A compilation of major shortcomings follows, and this aspect is examined in detail in the next chapters:

Wireless LANs

❏ In wireless local area networks there is no protection of the identity of mobile devices against eavesdropping on the radio interface since the globally unique MAC address of a network adapter is basically included in plaintext in each transmission frame.

❏ In GSM and UMTS networks, active attackers can retrieve the worldwide unique identification (called IMSI) of mobile devices so long as they are able to send forged signalling data over the radio interface to a device and receive messages from it. Furthermore, the operator of a 'visited' access network can monitor the movements of devices currently registered with it and relate this information to their unique identification. The contractual network operator (thus in a sense the home network operator) of a mobile device can in fact monitor all movements of the device, since it receives signalling requests and accounting data records from 'visited' networks for all devices that are contractually bound to it. In contrast, the current location of a mobile device remains hidden from its communication partners because this information is actually only accessible to network operators (and, upon court order, to prosecution services).

GSM and UMTS

❏ The approach for mobile Internet communication (Mobile IP) preferred by the IETF has glaring shortcomings in terms of location privacy. With 'standard Mobile IP' without extensions, infrastructure elements of the visited and the home network can track the movements of mobile devices. Furthermore, the identity of a mobile device is not even protected against eavesdropping during transmission over the radio interface. The same situation applies when mobile IP is used in connection with an AAA infrastructure. An even worse situation exists when mobile IP is used in combination with 'route optimisation': In this case even the arbitrary hosts to which routes are being optimised can track the current location of a mobile device by sending regular requests (e.g. 'ping').

Mobile IP

The basic problem with the design of mobile communication systems in terms of confidentiality of mobile device location is a conflict in goals. On one hand, each mobile device should be reachable for incoming communication requests; on the other hand, no (single) entity in the network should be able to track the current location of the device in the network.

Conflict in goals between reachability and untraceability

In recent years various fundamental approaches have been proposed to address this problem [MR99]:

❏ *Broadcast messages:* All messages are sent to all potential

Broadcast

receivers so that the location of the receivers does not have to be known. If the confidentiality of the messages being sent is also to be protected, then all messages are encrypted with a public key. In this case, all potential receivers have to decrypt all received messages to filter out the messages meant for them. It is obvious that this approach does not scale well for large networks or for high message loads.

Temporary pseudonyms

❏ *Temporary pseudonyms:* With this approach, all mobile devices do not use their actual identities but instead are contacted through pseudonyms that are regularly changed. An entity is necessary to map the current temporary pseudonym to the device to ensure that the mobile device can be reached. However, in principle, this entity can track and record the history of the temporary pseudonyms of a specific device.

Mix networks

❏ *Mix networks:* Messages are routed over various entities in the network with each entity only able to discover one part of the message route. These are called *communication mixes*.

Details about these approaches will be examined in the following sections.

14.2.1 Broadcast Communication

Explicit vs implicit addresses

Broadcast communication also distinguishes between *explicit* and *implicit addresses* for mobile devices. With explicit addresses (e.g. IP addresses) each entity that 'sees' a particular message is able to determine the addressed entity. In contrast, what distinguishes implicit addresses is that they do not identify a specific device or a specific location and instead only name an entity without any further meaning attributed to the name. Implicit addresses are usually selected randomly from a large address space to minimise the probability of random collisions.

Open and hidden implicit addresses

With implicit addresses, a further distinction is made between *open (visible)* and *hidden (invisible) implicit addresses*. The difference between the two versions is that each entity that sees multiple occurrences of the same visible address can check the equality of the visible address, whereas only the addressed entity can check invisible implicit addresses for equality.

Implementation of hidden implicit addresses

Hidden implicit addresses are implemented by using public key operations. The addressed entity A selects random number r_A and makes it known, together with its public key $+K_A$, to potential communication partners. Whenever entity B wants to send a message to A later on, it selects a fresh random number r_B and prepares the following hidden implicit address for A: $ImplAddr_A := \{r_B, r_A\}_{+K_A}$. If this address is sent in a broadcast message, A is the only entity that can determine that it was addressed in it because it is the only entity that can correctly decrypt the address.

14.2.2 Temporary Pseudonyms

The basic idea of using temporary pseudonyms to protect the confidentiality of the current location of a mobile device is that the current location of the device is no longer stored together with its identification ID_A but with a changing temporary pseudonym $P_A(t)$ instead. A trusted entity maps the identity ID_A to the current temporary pseudonym $P_A(t)$ but this entity does not have to know where the device is located at a particular time. *Basic idea of temporary pseudonyms*

When an incoming message is forwarded to the current location of device A, two procedures are needed to execute this operation: *Message forwarding*

1. First, the identity ID_A is mapped to the current temporary pseudonym $P_A(t)$ and the addressing information is adapted accordingly in the message.

2. During the second procedure, the message is forwarded to the current location of the device (for example, by checking the database for the current locations of all temporary pseudonyms).

In the case that the two functionalities are performed by independent entities, there is an assurance that no individual entity can prepare movement profiles of mobile devices. *Provisioning by independent entities*

What is important, of course, is that the entities that forward the message after mapping to the temporary pseudonym cannot discover the actual identity ID_A of the receiver from the message payload.

The communication mix explained below can be used to provide additional protection against attacks in which multiple entities in a network illegally exchange information in order to discover information about the movement profile of a mobile device.

14.2.3 Communication Mixes

The concept of communication mixes was invented in 1981 by Chaum for untraceable e-mail communication. A communication *Basic idea of communication mixes* mix uses the following measures to hide the communication relationships between sender and receiver:

Function of communication mixes

❏ It buffers incoming messages that are encrypted with its public key.

❏ It changes the appearance of the messages by decrypting them with its private key.

❏ It changes the sequences of the messages and always forwards them in batches ('batch processing').

Provided that a sufficiently high traffic load is processed, it can be ensured that even an attacker who can read all the incoming and outgoing messages of a communication mix is still not be able to relate incoming messages to outgoing ones. Therefore, no information can be deduced about the relationships between senders and receivers.

Cascading communication mixes

If, however, an attacker succeeds in compromising a communication mix, no further protection can be offered. This danger can be countered through an additional cascading of communication mixes. For example, in principle sender A can transmit message m to receiver B without being traced by including the two communication mixes $M1$ and $M2$:

$$A \rightarrow M_1: \ \left\{ r_1, M2, \left\{ r_2, B, \left\{ r_3, m \right\}_{+K_B} \right\}_{+K_{M2}} \right\}_{+K_{M1}}$$

$$M_1 \rightarrow M_2: \ \left\{ r_2, B, \left\{ r_3, m \right\}_{+K_B} \right\}_{+K_{M2}}$$

$$M_2 \rightarrow \ \ B: \ \left\{ r_3, m \right\}_{+K_B}$$

Required message volume

However, the security of this scheme depends on all communication mixes processing a sufficient volume of messages. The way to understand this concept is by imagining a network of communication mixes through which, in an extreme case, a single message is routed. Although the message is recoded in each mix, an attacker who can eavesdrop on all communication connections can follow the route of the message through all the cascaded mixes. Furthermore, all messages should be of the same length. The idea of cascaded networks has already been applied conceptually to mobile communication systems [MR99].

14.3   Summary

Mobile communication networks and wireless local area networks face the same threats as do fixed networks. The existence of wireless transmission links contributes considerably towards increasing the threat potential. It also makes it difficult to implement the necessary security services (e.g. key management for authentication in the access network area).

Increase in existing risk potential

Furthermore, the mobility of devices and their users introduces a new threat, which is the generation of movement profiles for individual devices. A range of theoretical concepts were proposed in the past to counter this threat: message transfer per broadcast, the use of temporary pseudonyms and the use of cascaded communication mixes that hide the connection between incoming and outgoing messages. So far these ideas have only been applied to mobile communication networks at a conceptual level.

New threat posed by generation of movement profiles

The following chapters in Part III are devoted to specific system examples. The subject of Chapter 15 is wireless local area networks with particular emphasis on the American standard IEEE 802.11. Chapter 16 describes and discusses the security services of mobile wide-area networks based on the GSM and UMTS standards. The final chapter examines the security aspects of mobile Internet communication with an emphasis on the IETF approach Mobile IP.

Overview of following chapters

The book [Sch03] is recommended to readers who are interested in a comprehensive introduction to the fundamental principles of the system examples discussed. The discussion in the following chapters only focuses on security aspects.

14.4   Supplemental Reading

[MR99] MÜLLER, G.; RANNENBERG, K. (Eds): *Multilateral Security in Communications*. Addison-Wesley-Longman, 1999

[Sch03] SCHILLER, J.: *Mobile Communications*. Second edition, Pearson Education, 2003

14.5   Questions

1. What are the new threats and difficulties in terms of communication and network security faced by mobile communication systems compared with fixed network communication?

2.   Why should the authenticity of a terminal be verified again if a mobile terminal executes a handover from one base station to another?

3.   How well does the method of hidden (invisible) implicit addresses scale as the number of transmitted messages increases?

4.   Discuss the usage possibilities of cascaded communication mixes for interactive applications such as Internet telephony or multimedia conferencing services.

15 Security in Wireless Local Area Networks

In recent years the American IEEE 802.11 standard has established itself as the key standard in the area of wireless local area networks [IEE97]. As the standard proposed by the *IEEE-802-LAN/MANS Standardisation Committee*, it defines medium access control (MAC) and the physical characteristics of a wireless local area network, also called a *wireless LAN* or just *WLAN* for short.

This chapter mainly examines the security properties of the standard. Until the media published a series of considerable shortcomings in the standard, most vendors of standard-conformant wireless network components were claiming that IEEE 802.11 'is as secure as a wired network'. The security flaws that occur are explained in detail in this chapter because this standard represents an excellent textbook example of a failed attempt at 'securing' a communication protocol. The chapter concludes with a description of the current efforts of the 802 committee to remove the shortcomings.

15.1 The IEEE 802.11 Standard for Wireless Local Area Networks

The 802.11 standard comprises four different options for the physical layer, each one supporting a range of different transmission speeds:

Physical layer

❏ *Frequency Hop Spread Spectrum (FHSS):* 1, 2, 5.5 and 11 Mbit/s

❏ *Direct Sequence Spread Spectrum (DSSS):* 1, 2, 5.5, 11 and 22 Mbit/s

❏ *Orthogonal Frequency Division Multiplexing (OFDM):* 6, 9, 12, 18, 24, 36, 48 and 54 Mbit/s

❏ *Baseband Infrared:* 1 and 2 Mbit/s

Medium access control sublayer

The MAC sublayer supports operation in infrastructure mode under the control of a base station as well as *ad hoc* communication between independent systems.

Figure 15.1
Components of an infrastructure network based on IEEE 802.11

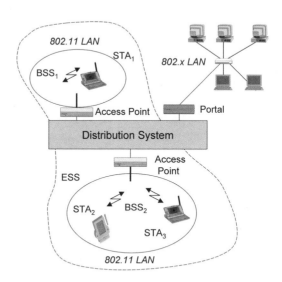

Figure 15.1 shows the components of a 802.11 WLAN when communication is in infrastructure mode:

Station

❏ A *station (STA)* is a system with access mechanisms to a wireless medium and radio contact to a base station.

Access point

❏ An *access point (AP)* is a (base) station with special protocol functions for medium access control and is usually incorporated into the wireless network as well as into a wired network (e.g. Fast Ethernet).

Basic service set

❏ A *basic service set (BSS)* is a group of stations using the same radio frequency.

Distribution system

❏ A *distribution system* is an interconnection network that combines multiple basic service sets into one logical network, also called an *extended service set (ESS)*.

Portal

❏ A *portal* is a bridge to other wired networks.

Figure 15.2 shows how communication is handled in *ad hoc* networks. In such networks, systems are only able to communicate with one another within a restricted radius as no infrastructure for connecting separate basic service sets exists.

Security functions of IEEE 802.11

The IEEE 802.11 standard uses the two security functions *entity authentication* and the protocol *Wired Equivalent Privacy*

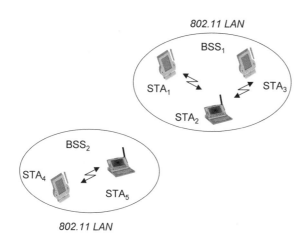

Figure 15.2
Ad hoc
*communication based
on IEEE 802.11*

(WEP), which provides the security services *confidentiality, data origin authentication and data integrity* as well as *access control* (in conjunction with layer management).

For these functions the WEP protocol uses the algorithms *RC4* (also see Section 3.4) and *cyclic redundancy check (CRC)*, which is actually an error-checking method and therefore does not meet the requirements outlined here. The two following sections explain the standard's security functions in detail.

15.2   Entity Authentication

The IEEE 802.11 standard contains a very rudimentary authentication function. Authentication is performed between stations (terminals) and base stations (access points) and optionally can also be performed between arbitrary terminals.

IEEE 802.11 provides two authentication schemes:

❑ *Open system authentication:* The standard itself [IEE01, Section 8.1.1] states: *'Essentially it is a null authentication algorithm'*. This algorithm therefore cannot offer any security regarding the identity of a communication partner.

Open system authentication

❑ *Shared key authentication:* With this scheme, a shared key is used to prove that a user belongs to the group that possesses this key. The standard does not regulate how the key is negotiated and distributed but it is explicitly assumed that the shared key was delivered to the stations over a secure channel independently of IEEE 802.11 [IEE01, Section

Shared key authentication

8.1.2]. Due to the lack of key management, many wireless LANs based on 802.11 are operated without authentication (or with 'open system authentication').

When shared key authentication is performed, one of the two stations acts as *requestor* and the other one as *responder*. The requestor starts the authentication process by asking the responder to conduct the following challenge-response exchanges:

$$A \rightarrow B: \ \left(Authentication, 1, ID_A\right)$$
$$B \rightarrow A: \ \left(Authentication, 2, r_B\right)$$
$$A \rightarrow B: \ \left\{Authentication, 3, r_B\right\}_{+K_{A,B}}$$
$$B \rightarrow A: \ \left(Authentication, 4, Successful\right)$$

This exchange authenticates the requestor to the responder. Two separate protocol runs — one per direction — are needed for mutual authentication. The encryption in the third protocol step is performed according to the WEP protocol (see below).

Insecure authentication protocol   Unfortunately, the authentication exchange described offers no security for the identity of the communication partner because it contains a serious[1] cryptographic error.

The protocol failure originates from a combination of two properties of the deployed mechanisms: a deficiency in the WEP protocol on how initialisation vectors needed to generate an RC4 keystream are selected, and the general characteristic of all OFB ciphers of supplying the corresponding keystream of a ciphered message with its plaintext in XOR operations.

Determining keystreams   Looking at the second and third steps in the protocol specification above, one notices that a potential attacker who is able to eavesdrop on a wireless medium knows the entire plaintext of the third authentication message ('$Authentication, 3, r_B$') and therefore can determine the keystream used to encrypt the message.

Reuse of initialisation vectors   As the WEP protocol also does not restrict the reuse of initialisation vectors and, consequently, the keystream (see also the section below), an attacker is able to use this keystream for his or her own authentication procedure without knowing the key $K_{A,B}$. This makes the 802.11 authentication exchange totally useless.

---

[1]This error is particularly serious because it is so easy to discover and should therefore have been noticed by the authors of the standard.

15.3 Wired Equivalent Privacy

The WEP protocol defines how encryption and integrity-protection should be performed to the user data of protocol data units of the MAC sublayer. The following sections explain the operations of WEP and its shortcomings.

15.3.1 Operation and Linearity of CRC

The WEP protocol provides integrity protection based on a cyclic redundancy check (CRC). This method uses an error check code that was actually developed as a statistical means of discovering random errors.

The mathematical basis of CRC consists of regarding bit strings as polynomials with the coefficients '0' and '1'. Therefore, message M is interpreted as polynomial $M(x)$ and polynomial arithmetic modulo 2 is used so that the addition and the subtraction correspond to a bitwise XOR operation. *Mathematical basis of CRC*

Prior to the computation of the CRC checksum of a message $M(x)$, sender and receiver agree on a *generator polynomial* $G(x)$ that is normally selected from a set of internationally standardised polynomials, i.e., it is specifically defined. Let n be the degree of the polynomial so that the length of the polynomial is $n + 1$.

Now consider the division of $M(x) \cdot 2^n$ by the generator polynomial $G(x)$ (note that in the polynomials considered the subtraction operation equals the addition equals a bitwise XOR operation): *Operation of CRC*

As $\dfrac{M(x) \cdot 2^n}{G(x)} = Q(x) + \dfrac{R(x)}{G(x)}$ is also $\dfrac{M(x) \cdot 2^n + R(x)}{G(x)} = Q(x)$

with R(x) being the remainder of $M(x) \cdot 2^n$ when divided by $G(x)$.

The remainder $R(x)$ after division by $G(x)$ is simply attached as a error checksum to message $M(x)$. The result $Q(x)$ of the division is not of further interest as the only thing considered in the verification of a received message $M(x) \cdot 2^n + R(x)$ is whether the remainder 0 results from division by $G(x)$.

Let us now consider the two messages M_1 and M_2 with the CRC values R_1 and R_2:

As $\dfrac{M_1(x) \cdot 2^n + R_1(x)}{G(x)}$ as well as $\dfrac{M_2(x) \cdot 2^n + R_2(x)}{G(x)}$ are both divisible with remainder 0, $\dfrac{M_1(x) \cdot 2^n + R_1(x) + M_2(x) \cdot 2^n + R_2(x)}{G(x)}$

also divides with remainder 0.

CRC is a linear function

What this proves is that CRC is a linear function, i.e., $CRC(M_1) + CRC(M_2) = CRC(M_1 + M_2)$. This property allows controlled modifications to messages and the corresponding CRC values and, therefore, makes CRC inappropriate for cryptographic purposes.

15.3.2　Operation of the WEP Protocol

Encryption using the RC4 algorithm

The WEP protocol uses the RC4 algorithm as a pseudo-random bit generator to generate a keystream that is always dependent on the key and an initialisation vector. A new 24-bit long initialisation vector IV is selected for each protected message M and concatenated with a shared key K_{BSS} that is normally known to all the stations of a basic service set. The resulting value serves as the input for the RC4 algorithm, which uses it to generate an arbitrarily long keystream (also see Section 3.4).

The CRC method is also used to compute a (presumed) *integrity check value (ICV)* that is appended to message M ('||'). The resulting message $(M \ || \ ICV)$ is XORed ('$\oplus$') with the keystream $RC4(IV \ || \ K_{BSS})$ and the initialisation vector in plaintext is placed in front of the resulting ciphertext. Figure 15.3 shows this process as a block diagram.

Figure 15.3
Block diagram of WEP encryption

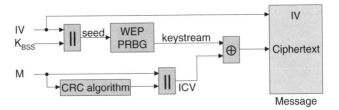

Self-synchronisation

As the initialisation vector is sent in plaintext with each message, each receiver that knows the key K_{BSS} can reproduce the appropriate plaintext. This process assures the property of *self-synchronisation* that is particularly important in wireless networks.

The decryption and integrity-verification process at the receiver essentially consists of the reverse processing sequence and is illustrated in Figure 15.4.

Security goals of WEP protocol

The WEP protocol was designed with the intention of assuring the following security goals:

❏ *Confidentiality:* Only those stations in possession of key K_{BSS} can read WEP-protected messages.

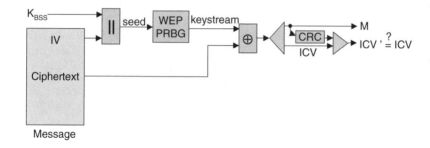

Figure 15.4
Block diagram of WEP decryption

❏ *Data origin authentication and integrity:* The receiver can detect malicious modifications to WEP-protected messages.

❏ *Access control in conjunction with layer management:* If preset as such in layer management, receivers only accept WEP-protected messages, which means that only those stations that know key K_{BSS} can send to these receivers.

The following section explains how WEP unfortunately does not achieve any of these goals.

15.3.3   Flaws in the WEP Protocol

The WEP protocol has security flaws in the following five areas:

1. Lack of key management.

2. Inadequate protection against messages being read by unauthorised parties.

3. Insecure data origin authentication and integrity protection.

4. Insufficient access control.

5. Key computation based on eavesdropped messages.

IEEE 802.11 does not define any specifications for *key management* and merely recommends the shared use of group keys. From a security view, shared group keys are considered to be very inadequate because the distribution of them is difficult, if not impossible, to control, and the fact that any member of a group can pretend to be another member of the group. Furthermore, due to the lack of key management, in practice group keys are very seldom if ever changed and in many cases the 802.11 security functions are even switched off.

Key management

Brute-force attacks on 40-bit long keys

Another point is that most WLAN products only support 40-bit long keys, and today this provides little protection against attackers who try out all possible keys to find the right one. Although some products do support 104-bit long keys,[2] they are often not interoperable with the 40-bit versions. Consequently, when a WLAN is introduced a decision has to be made at the outset about which devices are to be procured and the impact on the system if they are replaced in the future.

Messages read by unauthorised parties

Even if keys are properly distributed and sufficiently long in length, the WEP protocol does not *adequately protect messages from being read by unauthorised parties*. One reason is the reuse of the keystream [BGW01] that is a direct outcome of the short initialisation vectors and lack of key management. With WEP encryption, each message triggers a resynchronisation of the state of the pseudo-random bit generator by generating and adding a new 24-bit long initialisation vector IV in front of the group key K_{BSS}, and executing the initialisation procedures of the RC4 algorithm.

Reuse of initialisation vectors

Consider two messages M_1 and M_2 that were randomly encrypted using the same initialisation vector IV_1:

$$C_1 = P_1 \oplus RC4(IV_1, K_{BSS})$$
$$C_2 = P_2 \oplus RC4(IV_1, K_{BSS})$$

Therefore the following also holds:

$$C_1 \oplus C_2 = P_1 \oplus RC4(IV_1, K_{BSS}) \oplus P_2 \oplus RC4(IV_1, K_{BSS}) = P_1 \oplus P_2$$

Known-plaintext attack

If, for example, an attacker knows the values P_1 and C_1, he or she can compute plaintext P_2 from ciphertext C_2. The WEP protocol is therefore susceptible to known-plaintext attacks if initialisation vectors happen to be reused with the same key K_{BSS}.

This situation occurs relatively frequently in practice because many implementations choose the initialisation values poorly (for example, hardware is often initialised starting again with '0'). However, even if the initialisation values are selected in an ideal manner, the short 24-bit length remains a major problem. A busy WLAN base station, for example, exhausts its available IV space within half a day if transmission is at 11 Mbit/s.

---

[2]These products are often marketed with the misleading claim '128-bit encryption'. In reality, the key length is only 104 bits because 24 bits are used for the initialisation vector. From a cryptographic standpoint, 104-bit long keys are considered sufficiently secure to counter brute-force attacks under the assumptions of conventional computation models.

Due to the linearity of the CRC and the encryption function, an attacker can easily undermine WEP integrity-protection and data origin authentication. Let us consider a message sent from sender A to receiver B that is intercepted by attacker E:

Vulnerable data origin authentication

$$A \to B : (IV, C) \text{ with } C = RC4(IV, K_{BSS}) \oplus (M, CRC(M))$$

Even without knowing key K_{BSS}, attacker E can construct a ciphertext C' that is decrypted to a plaintext $(M', CRC(M'))$ with a valid CRC value. The attacker chooses an arbitrary message Δ and uses it to compute C' as follows:

$$
\begin{aligned}
C' &:= C \oplus (\Delta, CRC(\Delta)) \\
&= RC4(IV, K_{BSS}) \oplus (M, CRC(M)) \oplus (\Delta, CRC(\Delta)) \\
&= RC4(IV, K_{BSS}) \oplus (M \oplus \Delta, CRC(M) \oplus CRC(\Delta)) \\
&= RC4(IV, K_{BSS}) \oplus (M \oplus \Delta, CRC(M \oplus \Delta)) \\
&= RC4(IV, K_{BSS}) \oplus (M', CRC(M'))
\end{aligned}
$$

Because the attacker does not know the original message M, he or she also does not know the resulting message M'. However, what is known is that a '1' in a specific position in message Δ produces a 'flipped' bit in the corresponding position in M'. The attacker can therefore make controlled changes that will not be detected by the receiver.

Controlled modifications

The access control function of IEEE 802.11 can also easily be bypassed by attackers. All they need is a sufficiently long plaintext–ciphertext pair (M, C) (the IV value can be deduced from the eavesdropped message since it appears in front of the ciphertext). There are two main reasons why the access control function fails: the 'integrity verification function' CRC is not parameterised with a key and a lack of requirements for the reuse of initialisation values for RC4.

Inadequate access control

Because attacker E knows message M and ciphertext $C = RC4(IV, K_{BSS}) \oplus (M, CRC(M))$, she can compute the keystream $RC4(IV, K_{BSS})$ that matches the IV from it. If she wants to send her own message M' in the future, she uses the same initialisation value IV for transmission and computes the ciphertext of her message with the previously calculated and possibly shortened keystream:

$$E \to B : (IV, C') \text{ with } C' = RC4(IV, K_{BSS}) \oplus (M', CRC(M'))$$

As the reuse of arbitrary initialisation values is possible without triggering any alarms at the receiver, the attacker succeeds in creating valid messages that are accepted by any receivers in the WLAN.

Unauthorised use of network resources

This attack technique can be applied to make unauthorised use of network resources. For example, an attacker sends IP packets that are routed to the public Internet from the underlying network infrastructure of the WLAN. Even if the attacker cannot read the response packet as he does not know K_{BSS}, he or she will have already gained free access to the Internet and some 'useful applications'. These include the execution of *denial-of-service attacks*, where attackers are often not interested in the responses from receivers of their packets.

Computation of K_{BSS} using known plaintexts

An attack method that was published in August 2001 [FMS01] enables the key K_{BSS} to be retrieved after a sufficient volume of packets has been eavesdropped. This attack technique can retrieve a shared key in less than 15 minutes after the eavesdropping of around four to six million packets [SIR01]. It is based on the following properties of RC4 and the usage of RC4 in the WEP protocol:

1. RC4 is vulnerable to the deduction of individual key bits if a large number of messages are encrypted using a keystream computed from a variable initialisation value and a fixed key, and if the respective initialisation vector as well as the first two plaintext octets of each message are known.

2. With WEP, the initialisation value is transmitted in plaintext with each message.

3. The first two octets of each message can easily be guessed in WLANs because the first octets of IP packet are easily determined.

R. Rivest published the following comment on this attack technique [Riv01]: *'Those who are using the RC4-based WEP or WEP2 protocols to provide confidentiality of their 802.11 communications should consider these protocols to be broken [...]'*

Conclusion: WEP is totally insecure!

The overall conclusion is that the WEP protocol is totally insecure and users are urgently advised to use alternative security measures.

15.3.4 Alternative Security Measures

The two main ways to secure wireless networks based on IEEE 802.11 are either to use alternative security measures or to improve the standard itself. In the medium term, an improvement to the standard is the preferred option since the use of alternative measures would involve more of an effort in the planning, implementation and operation of a WLAN.

The IEEE therefore already has a task group involved in developing medium and longer term solutions to the security problems inherent in the standard. One aim of this two-step approach is to ensure that the medium-term solution can also be made available via software updates for existing systems already sold (if the vendor is willing and able to do so). The long-term solution is designed for future series of devices and therefore does not need to take into account any restrictions resulting from current hardware platforms.

IEEE task group to improve standard

So long as improvements to IEEE 802.11 are not yet standardised and available in corresponding software updates, users and operators of wireless LANs have no other choice but to use alternative measures to compensate for the security flaws found in 802.11.

Some minimal security measures should be implemented. If an Internet firewall exists, wireless LANs should always be placed outside the firewall or separated from the wired network infrastructure through a firewall installed expressly for that purpose. Devices that are connected through a WLAN should not be treated with more confidence than arbitrary external computers. Thus file systems should not be exported to computers in wireless LANs.

Alternative security measures

Furthermore, user data transmitted in a WLAN should be secured by security protocols above the MAC protocol layer. Part II of this book described a series of applicable protocols that are embedded in different layers of the communication architecture:

Security protocols above the MAC protocol layer

❏ Layer 2: IEEE 802.1x for access control, entity authentication and key management as well as PPP, PPTP and L2TP specifically for protecting user data from eavesdropping and modification (see also Chapter 10)

❏ Layer 3: IPSec (see also Chapter 11)

❏ Layer 4: SSL/TLS and SSH (see also Chapter 12)

A comparison of these alternatives and a study of the use of IPSec in wireless LANs with support of nomadic users can be found in [BS02]. One concern about using these protocols to secure WLAN communication is the possibility of interactions occurring with VPN configurations that already exist in the hosts. The preferred medium-term solution to the security problems in 802.11 networks is, therefore, one in which measures are directly integrated into the standard.

Improving the 802.11 The IEEE task group *802.11 Task Group I* is currently working
standard on the following two solutions (Status: August 2002):

❏ Medium term: Improve existing WEP protocol by changing
key schedulling for RC4 (*temporal key hash* [HW01]) and introduce an additional *message integrity check (MIC)* using
the function *'Michael'* [Fer02]).

❏ Long term: Specify a new security protocol that uses AES to
provide the security services confidentiality and data origin
authentication. According to a provisional draft [WHF02],
AES is be used for encryption in *counter mode*, with the AES
algorithm effectively deployed as a pseudo-random bit generator to compute the keystream. AES in CBC mode will be
used to compute MICs.

Upgrading existing At the moment it is very difficult to assess whether and when ex-
products isting products can be updated to include these enhancements.[3]
Although the alternative security measures described above can
be used independently of the support vendors provide for WLAN
products, it will be more time consuming and costly to incorporate
them into the implementation and operation of wireless LANs.

15.4 Summary

Due to the increased threat potential to radio transmission links,
special security measures are required for wireless local area networks. The IEEE 802.11 standard, which is the standard most
Security flaws in IEEE frequently used with such networks, does not provide adequate
802.11 protection due to the following security flaws:

❏ The lack of key management makes it difficult to use existing
security functions. The result in practice is that keys are
seldom changed if ever and security functions are often never
activated.

❏ Entity authentication and encryption are based on the use
of shared group keys, the distribution of which is difficult to
control. They also do not allow the protection of individual
data streams on a device-specific basis. The authentication
exchange is also totally useless as an attacker only needs

---

[3]It is possible that not all vendors will be able or prepared to provide
free updates.

to eavesdrop on a single authentication procedure to execute such an exchange.

❏ Many products only provide for 40-bit long keys, thus essentially offering no protection against brute-force attacks.

❏ The reuse of initialisation values for RC4 encryption makes it easy for known-plaintext attacks.

❏ The linearity of the 'integrity checking function' CRC allows controlled and undetected changes to be made to messages.

❏ The access control function can be bypassed an arbitrary number of times using a sufficiently long known plaintext–ciphertext pair.

❏ A weakness in the key schedulling for the RC4 algorithm enables the key to be computed using eavesdropped messages. A tool for executing this attack is freely available from the Internet.

IEEE-802.11 WLANs should therefore always be operated outside a firewall and secured through the use of additional security protocols (see also Chapters 10, 11 and 12). *Alternative security measures*

The Institute of Electrical and Electronics Engineers (IEEE) has set up a task group to improve the security shortcomings of the standard. The task group is working on medium and long-term solutions to the problems that exist. The medium-term solution would involve the use of software updates to enhance existing systems. *Improving the standard*

15.5 Supplemental Reading

[BGW01] BORISOV, N.; GOLDBERG, I.; WAGNER, D.: Intercepting Mobile Communications: The Insecurity of 802.11. In: *Proceedings of ACM MobiCom*, 2001. – http://www.cs.berkeley.edu/~daw/papers/wep-mob01.ps

[Fer02] FERGUSON, N.: *IEEE P802.11 Wireless LANs – Michael: an improved MIC for 802.11 WEP*. January 2002. – Institute of Electrical and Electronics Engineers (IEEE), Document IEEE 802.11-02/020r0

[FMS01] FLUHRER, S.; MANTIN, I.; SHAMIR, A.: Weaknesses in the Key Scheduling Algorithm of RC4. In: *Selected Areas in Cryptography, Lecture Notes in Computer Science* Vol. 2259, Springer, 2001, pp. 1–24

[HW01] HOUSLEY, R.; WHITING, D.: *IEEE P802.11 Wireless LANs – Temporal Key Hash*. December 2001. – Institute of Electrical and Electronics Engineers (IEEE), Document IEEE 802.11-01/550r3

[IEE97] INSTITUTE OF ELECTRICAL AND ELECTRONICS ENGINEERS (IEEE): *Wireless LAN Medium Access Control (MAC) and Physical Layer (PHY) Specifications*. The Institute of Electrical and Electronics Engineers (IEEE), IEEE Std 802.11-1997. 1997

[Riv01] RIVEST, R.: *RSA Security Response to Weaknesses in Key Scheduling Algorithm of RC4*. 2001. – http://www.rsa.com/rsalabs/technotes/wep.html

[SIR01] STUBBLEFIELD, A.; IOANNIDIS, J.; RUBIN, A. D.: *Using the Fluhrer, Mantin, and Shamir Attack to Break WEP*. August 2001. – AT&T Labs Technical Report TD-4ZCPZZ

[WHF02] WHITING, D.; HOUSLEY, R.; FERGUSON, N.: *IEEE P802.11 Wireless LANs – AES Encryption and Authentication Using CTR Mode and CBC-MAC*. May 2002. – Institute of Electrical and Electronics Engineers (IEEE), Document IEEE 802.11-02/001r2

15.6   Questions

1. Can you track down the errors in the shared key authentication protocol by conducting a formal analysis using GNY-Logics?

2. Why, with WEP, is the initialisation vector (IV) sent with each MAC-PDU instead of being implicitly extracted from the last MAC-PDU of the same sender to the same receiver?

3. Why is the combination of the integrity protection function (CRC), which is not parameterised with a key, and the ciphering algorithm (RC4), which is operated in OFB mode, not adequate for executing effective access control?

4. Give alternative options for implementing access control in wireless LANs.

5. Can use of IEEE 802.1x compensate for the security flaws in IEEE 802.11?

6. How can an attacker obtain possession of a plaintext–ciphertext pair if he or she knows that certain public visible IP addresses are used in the WLAN being attacked?

7. How would you evaluate WLANs based on IEEE 802.11 in terms of location privacy?

16 Security in Mobile Wide-Area Networks

The main difference between wireless local area networks and mobile wide-area networks are the functions for true mobility support that produce a range of qualitatively new aspects. This chapter explains the most important architectures currently being used in this area, *GSM* and *UMTS*, with special emphasis on their security functions.

16.1 GSM

The acronym GSM stands for the current leading standard for mobile telecommunications in the world. Standardised since the early 1980s by the *European Telecommunications Standards Institute (ETSI)*, it was originally specified as the acronym for the *Groupe Spéciale Mobile* in 1982 (at a time when the French language still played an important role in European telecommunications). GSM was renamed *Global System for Mobile Communication* during its triumphal period on the international stage in the 1990s.

GSM is a pan-European standard that was introduced in three phases in 1991, 1994 and 1996 by the European telecommunication administrations simultaneously. It enables the use across Europe of the services offered, referred to as *roaming*. Now GSM-based networks are being operated in more than 130 countries (in Africa, America, Asia, Australia and Europe). GSM allows 'true' mobile communication with no regional restrictions and supports the service types speech and data communication. GSM services can be used worldwide through a uniform address (an international telephone number).

Pan-European standard

Now being used in more than 130 countries

The description of GSM in this chapter mainly focuses on its security functions *confidentiality on the radio interface* and *access control and user authentication* (actually in the sense of device authentication). The GSM standards comprise the following security services [TG93, TS94]:

Security services of
GSM

❑ *Subscriber identity confidentiality:* Potential attackers should not be able to determine which user happens to be using given radio resources (data and signalling channels) simply by listening to the radio interface. Consequently, devices normally use a temporary pseudonym when they register. This provides certain protection against tracing a user's location through listening to the radio interface.

❑ *Subscriber identity authentication:* Prior to using the services of a network, each device must prove its identity in a simple authentication exchange.

❑ *Signalling information element confidentiality:* The confidentiality of signalling information exchanged between mobile devices and the access network is protected through the encryption of the signalling messages on the radio interface.

❑ *User data confidentiality:* Like signalling information, user data is encrypted on the radio interface.

The definitions of the security services listed above indicate that GSM measures mainly take into account passive attacks on the radio interface.

Architecture of a
GSM network

Before the operation of the security services can be explained, a brief overview is needed of the GSM network architecture. A detailed explanation of all components and their functions would take too much space in this chapter and readers are therefore referred to the relevant literature, e.g. [Sch03] for in-depth coverage.

Figure 16.1 presents an overview of the components of a GSM network. The acronyms of GSM terminology used in the illustration and other acronyms relevant to this chapter can be found in Table 16.1.

The signalling information and user data exchanged between mobile devices are encrypted in the area called *radio subsystem (RSS)*. To be more precise, encryption and decryption is performed by the mobile devices *(mobile station, MS)* and the base stations *(base transceiver station, BTS)*. The main tasks of the latter components include coding and decoding information transmitted on the radio interface as well as the appropriate modulation/demodulation and transmission.

A *base station controller (BSC)* normally controls multiple BTS. The BSC is also responsible for routing and forwarding calls between the *mobile switching centre (MSC)* and the base stations as well as allocating the resources used on the radio interface.

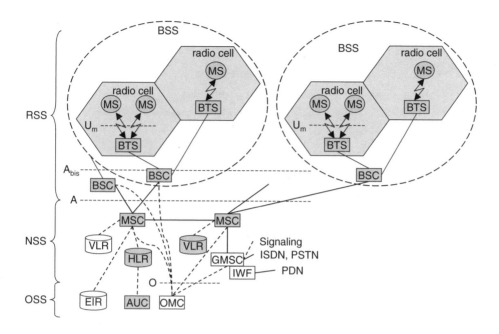

BSS

The two databases *home location register (HLR)* and *visited location register (VLR)* play a central role in the forwarding of incoming calls. The HLR contains information about the users administered by the network operator. This includes a user's identification *international mobile subscriber identification (IMSI)*, his/her mobile telephone number *mobile subscriber international ISDN number (MSISDN)* and other data on subscribed services. The HLR also stores the VLR number where a user is currently registered so that incoming calls can be routed to the appropriate network as well as the authentication vectors required for user authentication. These are explained below.

Figure 16.1
Architecture of a GSM network

The VLR is connected to one or multiple MSCs and registers the users that are currently located within a specific geographical zone, called *location area (LA)*. The VLR stores information about registered users similar to the HLR but with the addition of a *temporary mobile subscriber identity (TMSI)*, which is a temporary pseudonym to protect the identity of a user on the radio interface.

Figure 16.2 presents an overview of the principle involved in the authentication of mobile devices in GSM networks. After a mobile device inputs its IMSI to signal to an access network that it wants to use services of the network, the MSC obtains information required for the authentication from the HLR responsible for the device. The HLR does not generate this information itself;

Acronym	Meaning
AUC	Authentication Centre
BSC	Base Station Controller
BSS	Base Station Sub-System
BTS	Base Transceiver Station
EIR	Equipment Identity Register
GMSC	Gateway Mobile Switching Centre
HLR	Home Location Register
IMSI	International Mobile Subscriber Identity
ISDN	Integrated Services Digital Network
IWF	Interworking Function
LAI	Location Area Identifier
MS	Mobile Station (e.g. mobile telephone)
MSC	Mobile Switching Centre
MSISDN	Mobile Subscriber International ISDN Number
NSS	Network Subsystem
OMC	Operation and Management Centre
OSS	Operation Subsystem
PDN	Packet Data Network
PSTN	Public Switched Telecommunication Network
RSS	Radio Subsystem
TMSI	Temporary Mobile Subscriber Identity
VLR	Visitor Location Register

the *authentication centre (AUC)* that periodically or upon request sends the HLR a supply of authentication vectors for the device is responsible for this task.

Authentication vector   In GSM networks such an authentication vector consists of a triple $(r_i, SRES*_i, K_{BSC,MS:i})$. The r_i denotes a random number selected from the AUC and $SRES*_i$ is a *signed response*, i.e., $SRES*_i := A3(r_i, K_{MS})$, that the AUC computes by applying an algorithm called $A3$ to the random number r_i and the user-specific authentication key K_{MS}. After a mobile device has been successfully authenticated, the session key $K_{BSC,MS:i}$ is used to encrypt the signalling messages and the user data on the radio interface. This key is computed by the AUC from the random number r_i and

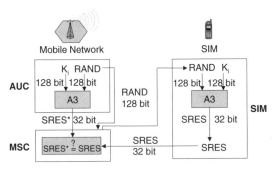

Figure 16.2
Authentication in GSM networks

the authentication key K_{MS} through use of the algorithm $A8$, i.e., $K_{BSC,MS:i} = A8(r_i, K_{MS})$.

The two algorithms A3 and A8 implement non-invertible functions. Like that of encryption algorithm A5, its specification is not publicly accessible and can only be obtained from system vendors, network operators or similar organisations by signing a confidentiality declaration from the GSM standardisation committee. In fact A3, A5 and A8 are only algorithm identifiers, and a range of different algorithms exist that are designed to meet all the properties specified in the GSM standards.

Cryptographic algorithms

The authentication key K_{MS} is stored additionally in the manipulation-safe *subscriber identity module (SIM)* of the mobile device. For authentication verification the access network (represented by the MSC and the BSC or the BTS) sends the mobile device the random number r_i of an authentication vector. The mobile device asks its SIM to use the random number and the stored key to compute the signed response that belongs to the random number and also to compute the session key: $SRES_i := A3(r_i, K_{MS})$; $K_{BSC,MS:i} := A8(r_i, K_{MS})$

Storage of authentication key in subscriber identity module (SIM)

The mobile device then responds to the access network with $SRES_i$. This $SRES_i$ denotes the response computed by the mobile device and only matches $SRES*_i$ if the correct key exists in the SIM. Upon receipt of this message, the access network verifies whether or not the response from the mobile device matches the expected response $SRES*_i$ (e.g. by the MSC as the GSM standard does not define which of the entities BTS, BSC, MSC or VLR performs this comparison). If the two values $SRES*_i$ and $SRES_i$ match, the authentication verification was successful.

After this verification the encryption is activated on the radio interface and the access network notifies the mobile device of the *location area identifier* (LAI_{VLR1}) for the current region and the new temporary pseudonym $TMSI_{MS:n}$. Thus, the authentication exchange can be summarised as follows:

1. $MS \rightarrow MSC : (IMSI_{MS})$

2. $MSC \rightarrow HLR : (IMSI_{MS})$

3. $HLR \rightarrow MSC : (IMSI_{MS}, r_{[i,j]}, SRES*_{[i,j]}, K_{BSC,MS:[i,j]})$

4. $MSC \rightarrow MS : (r_i)$

5. $MS \rightarrow MSC : (SRES_i)$

6. $MSC \rightarrow MS : (\{LAI_{VLR1}, TMSI_{MS:n}\}_{K_{BSC,MS:i}})$

The HLR normally does not transfer one but a series of authentication vectors (5 is considered a normal number and is shown as the notation '$[i, j]$' above) to the MSC to avoid the need for contact between the access network and the HLR for every authentication procedure. After successful authentication of the mobile device, all other protocol data units transmitted over the radio interface are encrypted and thus protected from potential eavesdropping attacks.

Handover within a location area   The procedure described so far is used when a mobile device 'initially' registers in a specific access network, e.g. after the device is switched on. If the device has already registered once in the access network, the IMSI is not required for a new authentication. The authentication is performed according to the following scheme:

1. $MS \rightarrow MSC : (LAI_{VLR1}, TMSI_{MS:n})$

2. $MSC \rightarrow MS : (r_i)$

3. $MS \rightarrow MSC : (SRES_i)$

4. $MSC \rightarrow MS : (\{LAI_{VLR1}, TMSI_{MS:n+1}\}_{K_{BSC,MS:i}})$

After each successful authentication procedure a new temporary pseudonym is allocated to the mobile device. Therefore, an attacker who merely listens to the radio interface cannot find out which devices are active in the respective radio cell nor even allocate consecutive authentication requests (and thus service uses) to individual devices.

The location area identifier sent with each re-authentication request allows devices coming from other access network areas to be identified. If the old and the new region happen to fall under the control of the same network operator, a handover can be executed with a change of the VLR. This takes place according to the following scheme:

Handover between two location areas

1. $MS \rightarrow VLR2 : (LAI_{VLR1}, TMSI_{MS:n})$

2. $VLR2 \rightarrow VLR1 : (LAI_{VLR1}, TMSI_{MS:n})$

3. $VLR1 \rightarrow VLR2 : (TMSI_{MS:n}, IMSI_{MS}, K_{BSC,MS:n}, r_{[i,j]},$
   $\qquad SRES*_{[i,j]}, K_{BSC,MS:[i,j]})$

4. $VLR2 \rightarrow MS : (r_i)$

5. $MS \rightarrow VLR2 : (SRES_i)$

6. $VLR2 \rightarrow MS : (\{LAI_{VLR2}, TMSI_{MS:n+1}\}_{K_{BSC,MS:i}})$

Upon receipt of the first message, the $VLR2$ (or the BSC or MSC processing the message) determines that the LAI contained in the message falls under the responsibility of a different VLR. It then requests information about the device identified by $TMSI_{MS:n}$ from the appropriate VLR. The queried $VLR1$ responds with the IMSI of the mobile device and the remaining authentication vectors. Afterwards the authentication exchange between VLR2 and the mobile device can continue as explained above.

If the $VLR1$ has no other unused authentication vectors for the mobile device, in other words, an allocation of a TMSI to the IMSI of the device is no longer possible (e.g. if a device has been switched off for a long time), $VLR2$ explicitly asks the mobile device for its IMSI in order to execute an initial authentication procedure (see above). The mobile device cannot encrypt the IMSI, which means that a potential attacker has the possibility of listening to the device's IMSI. An attacker can also exploit this weakness for an explicit retrieval of the IMSIs of mobile devices. Because mobile devices do not have the possibility of verifying the entities of an access network, they are unable to detect or prevent this kind of attack.

Recourse to an initial authentication exchange

To summarise GSM security mechanisms, GSM authentication is based on a challenge-response procedure that only establishes the identity of mobile devices to the access network. Mobile devices themselves have no possibility of verifying the identity of the entities of an access network. The periodic allocation of temporary identities at least enables the identity of these mobile devices to be

Achieved security properties

protected from passive attackers on the radio interface. However, active attackers can bypass this protection by explicitly sending requests to mobile devices.

The encryption of data units after the successful authentication of a device provides protection from attackers who listen to the radio interface. However, within the GSM network and also at the gateway into other network areas (e.g. ISDN) data units are exchanged unencrypted and no further security measures are taken (e.g. no authentication of signalling entities). The underlying model for the security mechanisms of GSM basically implies that all network operators can trust one another and that the signalling network is secure from attackers. Concluding, GSM networks are designed to be only as secure as conventional (non-secured) telecommunication networks.

16.2   UMTS Release '99

The acronym *UMTS* stands for *Universal Mobile Telecommunications System* and denotes a third-generation mobile communication system that was mostly developed in Europe. As such, it is part of the worldwide standardisation under the umbrella *International Mobile Telecommunications (IMT)*, which is coordinated by the *International Telecommunication Union (ITU)*. UMTS is thus a technology for IMT-2000 that is standardised within the framework of the ITU.

The telecommunications industry as well as European politics have great hopes for UMTS, which should play a key role in establishing a mass market for wireless multimedia communication with high service quality and continue the success of European GSM technology into the age of the mobile information society.

For space reasons, this chapter only looks at the security concepts of UMTS and readers are referred to [Les03] for a detailed introduction to the technology. This section specifically discusses the version *Release '99* with particular emphasis on the important aspects of device authentication and key negotiation.

Security services of UMTS Release '99   UMTS Release '99 provides the following security services [3GP02]:

User identity confidentiality   ❏ *User identity confidentiality* to pursue the following specific security goals:

❏ User identity confidentiality: The permanent identity of a user (his or her IMSI) cannot be eavesdropped on the radio interface.

❏ User location confidentiality: The current presence of a user in a particular geographic region cannot be discovered through eavesdropping on the radio interface.

❏ User untraceability: Potential attackers cannot eavesdrop on the radio interface to determine whether multiple services are being provided simultaneously to a particular user.

❏ *Entity authentication* will be available in two versions:

Entity authentication

❏ User authentication should prove the identity of a user (or his or her device) to the access network currently being used.

❏ Network authentication should enable a user to verify that he or she is connected to the access network authorised by the respective home network to offer services (this incorporates the currentness of this authorisation).

❏ *Confidentiality* with the following specific security goals or properties:

Confidentiality of user data

❏ Algorithm negotiation (cipher algorithm agreement) on a manipulation-safe basis between a mobile device and the access network it aims to connect to.

❏ Key negotiation (cipher key agreement) for the session keys used by a mobile device and the access network.

❏ Confidentiality of user data to ensure that user data cannot be eavesdropped on the radio interface.

❏ Confidentiality of signalling information to ensure that it neither can be eavesdropped on the radio interface.

❏ *Data integrity*, which pursues the following goals:

Data integrity

❏ algorithm negotiation (integrity algorithm agreement);

❏ key negotiation (integrity key agreement);

❏ data origin authentication and data integrity of signalling information so that each entity receiving signalling information (mobile device or access network entity) can verify prior to a message being processed whether it actually originates from the entity indicated as the sender and was not manipulated on the radio interface during transmission.

Like GSM, UMTS also mainly considers the security needs of the network operator

What is clear from the enumeration above is that the security mechanisms of UMTS Release '99 mainly focus on eavesdropping attacks. Aside from protecting authentication exchanges (see below) from active attackers on the air interface, it only offers integrity protection for signalling data. This means that the system mainly considers those security aspects that are fundamentally important to network operators. However, unlike GSM, mobile devices can gain assurance of the 'trustworthiness' of the current access network during an authentication exchange.

Certain acronyms are commonly used in conjunction with the authentication and key negotiation functions of UMTS explained below. These acronyms are listed in Table 16.2.

Table 16.2
Common abbreviations used with UMTS authentication

Acronym	Meaning
AK	Anonymity Key
AMF	Authentication Management Field
AUTN	Authentication Token
AV	Authentication Vector
CK	Confidentiality Key
GPRS	General Packet Radio Service
HE	Home Environment
IK	Integrity Key
RAND	Random Challenge
SGSN	Serving GPRS Support Node
SN	Serving Network
SQN	Sequence Number
USIM	User Services Identity Module
XRES	Expected Response

UMTS authentication

Figure 16.3 shows a UMTS authentication exchange. As with GSM, three principal actors can be identified: (1) the mobile device (mobile station, MS); (2) the network (serving network, SN) currently being used by the mobile device and represented by one of the two entities VLR or SGSN (= serving GPRS support node, which provides mobility functions for data services similar to those provided by a VLR for voice service) and (3) the home network of the mobile device, represented by one of the two entities HLR or HE (home environment).

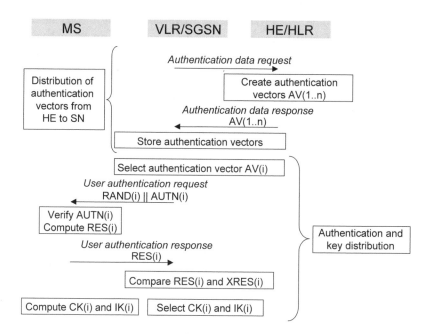

Figure 16.3
Overview of authentication exchange with UMTS Release '99

UMTS authentication is divided into the two phases *transport of authentication vectors* and the actual *authentication and key negotiation*. As is the case with GSM, the first phase is triggered when a mobile device wants to use the services in a particular network and the access network establishes that it has to request authentication vectors (AV) for the device. Once these vectors are received and stored, actual authentication and key negotiation can be performed for a specific connection. This procedure, called *authentication and key agreement (AKA)* in UMTS terminology, follows a pattern similar to GSM but with some specific differences.

Phases of UMTS authentication

The responsible entity of the access network sends a random number $RAND(i)$ (according to UMTS terminology) and an authentication token $AUTN(i)$ to the mobile device. The mobile device computes a response $RES(i)$ from the random number and the key K stored in its USIM (user services identity module) and using $AUTN(i)$ verifies whether the access network entity was authorised by its home network (see below). If this verification is successful, the mobile device sends the response $RES(i)$ to the ac-

UMTS authentication exchange

cess network. The mobile device then has its USIM compute the two session keys $CK(i)$ (confidentiality key) and $IK(i)$ (integrity key) from the random number and key K, which is stored in the USIM.

The access network compares the response received with the expected response $XRES(i)$. If the two values do not agree, the access network discontinues the communication. On the other hand, if the values agree, the access network extracts the two session keys $CK(i)$ and $IK(i)$ from the authentication vector $AV(i)$ and activates integrity protection and encryption on the radio interface.

Authenticating access network to the mobile device

One important innovation offered by UMTS but not available with GMS is the possibility given to a mobile device to verify whether its communication partner in an authentication exchange actually has the information generated by its home network, and whether this information is current. Section 7.2 explained that there are basically two possibilities for verifying the currentness of messages during an exchange: random numbers and time stamps (note that time stamps do not necessarily have to contain absolute times and instead can also be sequence numbers). With UMTS Release '99 the mobile device checks the currentness of the network's answer message during an authentication exchange using sequence numbers, whereas the currentness check by the network is performed on the basis of random numbers.

Generating authentication vectors

Figure 16.4 presents an overview of how authentication vectors are generated. The HE/HLR or AUC (authentication center) first generates a new random number $RAND$ and increments the sequence number SQN that is synchronised between the mobile device and the HE/AUC. These two numbers are used with five standardised functions $f1_K$, ..., $f5_K$ that are always parameterised with the key K of the mobile device to compute the following values:

❏ message authentication code
$MAC := f1_K(SQN\|RAND\|AMF)$

❏ expected response $XRES := f2_K(RAND)$

❏ confidentiality key $CK := f3_K(RAND)$

❏ integrity key $IK := f4_K(RAND)$

❏ anonymity key $AK := f5_K(RAND)$

The authentication token is constructed from these values according to the following rule:

$$AUTN := SQN \oplus AK \| AMF \| MAC$$

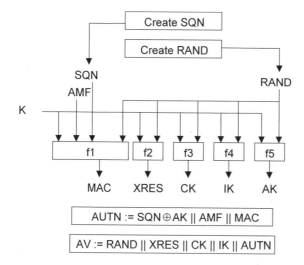

Figure 16.4
Generating authentication vectors with UMTS Release '99

Encryption of sequence numbers

The task of the anonymity key is to protect sequence numbers from potential eavesdropping attacks on the radio interface during transmission. Otherwise the sequence numbers could be used to conclude that consecutive authentication procedures are being executed by one and the same mobile device. AMF denotes the *authentication and key management field*. The UMTS standards do not include any compulsory specifications regarding its tasks and its use is on an operator-specific basis.

The authentication vector comprises the following components:

$$AV := RAND \| XRES \| CK \| IK \| AUTN$$

Figure 16.5 shows the processing procedures for an authentication exchange in a mobile device. Upon receipt of the random number *RAND* and the authentication token *AUTN*, the device computes the anonymity key $AK := f5_K(RAND)$. It then uses this key to extract the sequence number from the authentication token.

Figure 16.5
*Client-side processing
with Release '99
authentication*

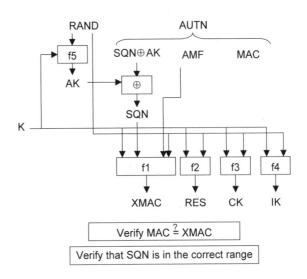

Protection against
replay attacks
At this point the sequence number can already be used to perform a currentness test of the received message, whereas the authenticity check first requires a verification of the MAC. The device computes the expected MAC $XMAC := f1_K(SQN\|RAND\|AMF)$ and compares it to the received value MAC. If both values agree and the sequence number is sufficiently current (the UMTS standard recommends that the sequence number should originate from a range of 32 current numbers), the device assumes that the current access network has received the tuple $(RAND, AUTN)$ from its home network and therefore has authorisation from it to provide services. If the sequence number does not fall within the correct range but the MAC is correct, the mobile device sends the message *synchronisation failure* with an appropriate parameter for resynchronising the sequence numbers.

After this verification the mobile device computes its response $RES := f2_K(RAND)$, the confidentiality key $CK := f3_K(RAND)$ and the integrity key $IK := f4_K(RAND)$ and sends the value RES to the access network.

As mentioned earlier, the peer entity in the access network then verifies whether the received value RES agrees with the expected value $XRES$. If this verification is successful, integrity protection and encryption are activated on the radio interface and a new temporary identity $(TMSI_n, LAI_n)$ is forwarded to the mobile device.

16.3 Summary

The two international standards GSM and UMTS are of central importance in the area of radio-based wide-area networks. Their security properties share a range of basic similarities.

Both architectures are based on the assumption that network operators within line-switched network areas can trust each other and no attacks will be made on transmitted data. Both architectures therefore deploy security measures that mainly apply to user and device authentication and transmission security over the radio interface.

With GSM and UMTS the different network operators have to trust one another

Device authentication for both architectures is also based on similar principles. A trustworthy entity AUC generates authentication vectors that contain challenge-response pairs as well as session keys. These vectors are provided to a visited access network upon request, although the transmission of these vectors is not secured in the signalling network. The signalling network always selects one of these vectors and sends the contained random number to the mobile device being authenticated. The device reacts with a response which it computes from the random number and the key stored in its security module. The session key(s) are generated in the same way.

Device authentication

Both architectures deploy temporary pseudonyms to conceal the identity of mobile devices on the radio interface during authentication. However, this measure does not protect against active attackers as both protocols explicitly allow access networks to request the permanent identity of a mobile device at any time. Furthermore, with both GSM and UMTS the confidentiality of transmitted data is only protected on the radio interface.

Use of temporary pseudonyms

UMTS additionally offers integrity protection for signalling information on the radio interface and gives mobile devices the possibility to verify the authenticity, as well as the currentness, of the data sent by an access network during an authentication exchange.

Innovations of UMTS vis-à-vis GSM

In summary, both systems were designed with the aim of providing a level of security comparable to a fixed network without any explicit security measures.

16.4 Supplemental Reading

[3GP02] 3GPP: *3G Security: Security Architecture (Release 1999)*. 3rd Generation Partnership Project, Technical Specifi-

cation Group Services and System Aspects, 3GPP TS
33.102, V3.12.0. June 2002

[Les03] LESCUYER, P.: *UMTS: Its Origins, Architecture and the
Standard.* Springer, 2003

[Sch03] SCHILLER, J.: *Mobile Communications.* Second edition,
Pearson Education, 2003

[TG93]   TC-GSM, ETSI: *GSM Security Aspects (GSM 02.09).* European Telecommunications Standards Institute (ETSI),
Recommendation GSM 02.09, Version 3.1.0, June 1993

[TS94]   TC-SMG, ETSI: *European Digital Cellular Telecommunications System (Phase 2): Security Related Network
Functions (GSM 03.20).* European Telecommunications
Standards Institute (ETSI), ETS 300 534, September
1994

16.5   Questions

1. Which new security services does UMTS offer compared to
    GSM?

2. What is the purpose of the location area identifier of GSM?

3. For what purpose does GSM use a TMSI? Why must a mobile
    device send an IMSI upon request?

4. How does a mobile terminal retrieve the current session key
    for communication with the base station?

5. Why is communication with GSM not encrypted end-to-end
    and only on the radio interface?

6. What would be the disadvantage if random numbers were
    also used as part of the currentness test performed by a mo-
    bile device during an authentication exchange under UMTS
    Release '99?

7. What is the purpose of the anonymity key with UMTS
    authentication?

8. What is the effect of the loss of synchronisation of the
    sequence numbers used by the home network to prove the
    currentness of its messages to a mobile device?

9. Assume that the temporary identity of UMTS is also used for the home network so that a visited network cannot discover the IMSI. What potential difficulties could this cause in actual network operations? Which category of cryptographic schemes would have to be used in such a case to provide secure identity of the mobile device?

10. Formulate the authentication protocol of UMTS Release '99 according to the notation mainly used in this book (see also Chapter 7 and particularly Table 7.1).

17 Security of Mobile Internet Communication

Due to the increasing use of data and multimedia services, there is also now a need to provide these services to mobile users on the basis of the dominant TCP/IP protocol suite. This chapter introduces *Mobile IP*, the currently preferred approach of the IETF to meet this need. Again the focus is particularly on security aspects and readers are referred to the appropriate technical books for a general introduction to Mobile IP, e.g. [Sch03, Section 9.1].

17.1 Mobile IP

Any consideration of the TCP/IP protocol suite for providing mobile communication has to take into account that the Internet was basically designed for stationary systems. Routing and packet forwarding is based on an IP destination address given in the packet. IP addresses are structured hierarchically and a network prefix determines the IP subnetwork. An IP address therefore has two functions:

The Internet was designed for stationary systems

❏ it *names* an end system;

❏ it *localises* the end system.

Double function of an IP address

Consequently, a change of the physical subnetwork to which a system is attached also implies a change of the IP address if the property of IP addresses *topological correctness* is to be maintained.

The theoretically conceivable approaches of *introducing specific paths for individual end systems* and *changing the IP address when a system moves from one subnetwork to another* do not scale with an increase in the number of end systems and frequent location changes. They create considerable problems with identifying individual end systems and have a negative effect on other protocol functions (e.g. breaking off TCP connections when an IP address is changed).

Figure 17.1
Overview of entities and data flow with Mobile IP

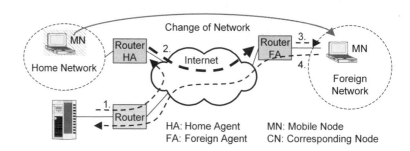

Basic idea of Mobile IP   The basic idea behind *Mobile IP* [Per96, Per02], the currently preferred approach of the IETF, consists of the introduction of an indirection stage based on a supplemental temporary IP address. Figure 17.1 presents an overview of the architecture of mobility support with Mobile IP. It introduces the following entities and concepts:

Mobile node (MN)   ❏ *Mobile node (MN):* an end system that can change its point of connection to a network without changing its IP address; typically a mobile end system.

Corresponding Node (CN)   ❏ *Corresponding node (CN):* an arbitrary system in the Internet with which the MN communicates. A CN does not need to know about the mobility of an MN and also requires no mobility support capabilities.

Home agent (HA)   ❏ *Home agent (HA):* a system in the home network of the mobile node, typically a router. The HA registers the current location of a mobile node, encapsulates incoming IP packets for the mobile node and routes them to the care-of-address (see below).

Foreign agent (FA)   ❏ *Foreign (agent (FA):* a system in the network in which a mobile node is currently operating and typically also a router. The FA forwards the encapsulated packets received from the HA to the mobile node, and is usually also the default-gateway for the mobile node. The foreign agent also offers the care-of-address for the mobile node.

Care-of-address (COA)   ❏ *Care-of-address (COA):* the address of the current tunnel end point for packets forwarded to the mobile nodes. From the IP point of view, the COA is the current address and current location area of the mobile node and can be selected by the mobile node using an FA or the Dynamic Host Configuration Protocol (DHCP).

IP packets addressed to mobile nodes are forwarded according to the following procedure:

Packet forwarding to MN

1. The corresponding node sends its packets to the permanent address of the mobile node. As is customary in the Internet, the packets are routed to the home network of the MN and intercepted there by a home agent.

2. The home agent encapsulates the packets (see below) and forwards them to the care-of-address. The care-of-address either addresses the foreign agent (as assumed in Figure 17.1) or is temporarily allocated to the mobile node (Mobile IP supports both versions).

3. If the COA addresses the FA, then the FA decapsulates the packet and forwards it to the mobile node. To do so, the FA must be able to address the packets directly to the mobile node on layer 2 because the IP destination address will otherwise not be topologically correct in the foreign network. The FA therefore must be in the same layer 2 subnetwork as the MN. If the COA is allocated temporarily to the MN, the MN decapsulates the packets itself.

With standard-Mobile IP the mobile node sends response packets directly to the CN, entering its permanent IP address as the source address. Thus, the mobile node is sending packets with topologically incorrect source addresses in the foreign network.

Packets of MN carry topologically incorrect source addresses

The procedure described has the advantage that it is largely transparent for end systems and that no modification is required to conventional gateway systems.

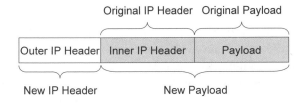

Original IP Header   Original Payload

Outer IP Header | Inner IP Header | Payload

New IP Header   New Payload

Figure 17.2
Packet encapsulation for data forwarding with Mobile IP

Packets directed to the permanent IP address of a mobile node are forwarded in the form of encapsulated packets. Figure 17.2 shows the fundamental structure of such packets. A new packet

Packet encapsulation

header is attached to the front of the original IP packet with a packet header and payload, so that the original packet is completely regarded as payload for the new packet.

Mobile IP supports different forms of packet encapsulation: IP-in-IP encapsulation, minimal encapsulation and encapsulation with GRE (see also Section 10.3).

Figure 17.3
Packet structure of
IP-in-IP encapsulation

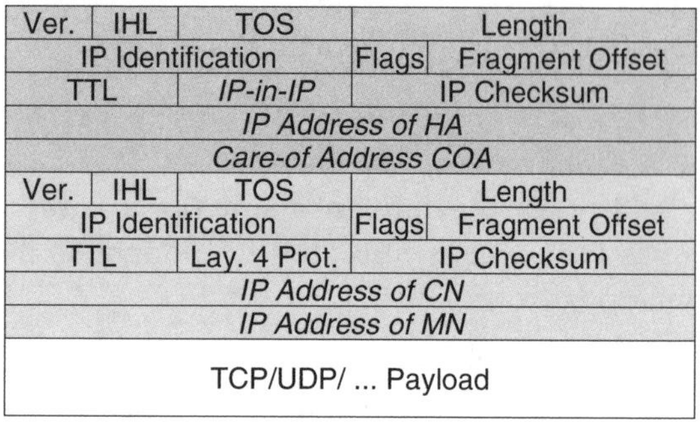

Ver.	IHL	TOS	Length	
IP Identification			Flags	Fragment Offset
TTL		IP-in-IP	IP Checksum	
IP Address of HA				
Care-of Address COA				
Ver.	IHL	TOS	Length	
IP Identification			Flags	Fragment Offset
TTL		Lay. 4 Prot.	IP Checksum	
IP Address of CN				
IP Address of MN				
TCP/UDP/ ... Payload				

IP-in-IP encapsulation   The packet structure for IP-in-IP encapsulation is shown in Figure 17.3. The address of the home agent is entered as the source address in the outer IP packet header and the care-of-address is entered as the destination address. 'IP-in-IP' is noted as the protocol identification because an IP packet follows as the payload. In the inner packet header the address of the CN is left unchanged as the source address and the address of the mobile node is left unchanged as the destination address.

Network integration   Before data packets can be exchanged in a tunnel between the
of a mobile node   home agent and the current care-of-address of the mobile node, the mobile node must first be integrated into the access network. In Mobile IP a distinction is made between the following three functionalities:

Agent advertisement   ❏ *Agent advertisement:* Home and foreign agents periodically send agent advertisement messages to the subnetworks for which they are providing services. Mobile nodes receiving these messages can then detect whether they are currently in their home network or in an 'outside' network, in other words, whether they have to change the foreign network (e.g. if no more advertisements are being received from the previous access network). A mobile node can further extract the care-of-addresses offered by the foreign agent.

❏ *Registration:* If a mobile node detects that its access network has changed ('movement detection'), it extracts a care-
of-address from the advertisement message it received and
requests that the address be registered. An overview of this
process in presented in Figure 17.4. The mobile node sends
a registration message with the previously determined care-
of-address to the foreign agent who forwards it to the home
agent. The process is secured through appending and checking message authentication codes (MAC, see below). The mobile node signals its home agent when returning to the home
network.

Registration

❏ *Local advertisement of the permanent IP address of a mobile
node in the home network:* After a home agent receives a
registration message for one of the mobile nodes it admin-
isters and has verified its authenticity, it advertises in the
local home network of the mobile node that packets for the
permanent IP address of the mobile node must in future be
sent to the layer 2 address of the home agent ('ARP advertisement'). From this point onwards the home agent receives all
IP packets addressed to the mobile node, encapsulates them
and forwards them to the care-of-address. When the mobile
node returns to its home network, it informs the home agent
accordingly and advertises its own layer 2 address as the destination for IP packets sent to it.

ARP advertisement

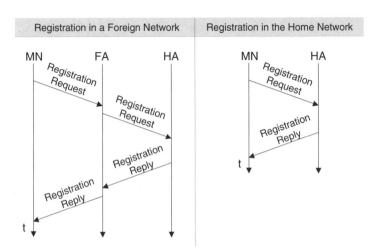

Figure 17.4
*Registration with
Mobile IP*

Triangular routing   The downside of the original Mobile IP [Per96] described above is called *triangular routing*: Packets to a mobile node are always diverted to the home agent on their way to the mobile node, whereas the mobile node can respond directly. The result is higher latency in one direction and a higher overall network load. This proves to be an unnecessary disadvantage, particularly when a mobile node is operating in the immediate vicinity of the corresponding node.

Route optimisation   One solution to this problem is the route optimisation for Mobile IP recommended in [PJ00]. A *binding update (BU)* notifies the corresponding node of the current care-of-address (and therefore also the position) of the mobile node. The CN, which needs basic mobility support functions to support this option, can then encapsulate its packets itself and send them directly to the MN.

However, this optimisation is more time-consuming if a mobile node has to change its care-of-address due to a change of foreign agent. In this case, not only the home agent but also all corresponding nodes must be informed of the location change.

Packets that are already in transit to the old COA at the time the mobile node changes location can be lost. The new FA can prevent this from happening by informing the old FA about the location change so that the old FA can forward packets still arriving for the mobile node to the new FA. Furthermore, the FA can also release any resources that might be reserved for the mobile node (storage, bandwidth, etc.).

Figure 17.5
Registration messages
with binding updates

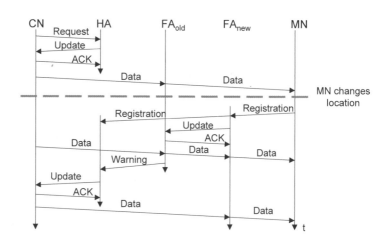

Signalling and route   Figure 17.5 shows the signalling messages exchanged in connec-
optimisation   tion with route optimisation. The first IP packet sent from CN to MN is routed to the home network and interpreted there by the

home agent as a *request*. The home agent responds with a *binding update* that contains the current COA of the MN and is acknowledged per *ACK* by the CN. From that point onward the CN encapsulates all data packets destined to the MN itself and sends them directly to the COA of the MN (shown as FA_{old} in the illustration).

When the MN changes the foreign agent, it sends a registration message through the new FA_{new} to its home agent. After this message is processed and forwarded, the new foreign agent sends an *update* message to inform the old foreign agent that acknowledges the message with an *ACK*. The old foreign agent then forwards all messages it is still receiving for the mobile node to the new foreign agent. In addition, it uses a *warning* message to inform the home agent about each CN that is still sending its packets to the old COA of the MN. With a new *update* message the home agent then informs each CN of the current COA. The CN acknowledges the receipt of each *update* message and subsequently sends its packets directly to the new COA.

Another source of problems with the original Mobile IP is that mobile nodes use their permanent IP addresses when sending in foreign access networks. However, many routers now only forward packets with topologically correct source addresses because, aside from Mobile IP, topologically incorrect addresses can be used for a variety of different attacks (see also Chapter 13).

Reverse tunneling

Consequently, RFC 2344, which was updated by RFC 3024 [Mon98, Mon01], specified the optional use of *reverse tunnels* for packets sent by mobile nodes. With reverse tunnels, the foreign agent encapsulates the IP packets sent by a mobile node and forwards them to the home agent. The home agent decapsulates the IP packets and forwards them to the addressed receiver. This extension is relatively easy to implement and is compatible with existing implementations of Mobile IP without them having knowledge of it. It also solves the problem of TTL values that are too low in packets sent by mobile nodes.

However, the use of reverse tunnels can also result in problems. For example, it is important to ensure that Mobile IP tunnels are not misused to allow IP packets into a subnetwork even though they would otherwise have been rejected by the Internet firewall of the subnetwork. In addition, data transfer is less efficient because now packets of the mobile node are also routed via the home network when they are forwarded to the addressed node ('double triangular routing').

Potential misuse of Mobile IP tunnels

17.2 Security Aspects of Mobile IP

Original security aspects considered

A range of security aspects were identified when original Mobile IP was specified in RFC 2002 [Per96]:

❑ Because data traffic for a mobile node can be re-routed by performing a Mobile IP registration procedure, the potential for misuse would be considerable if effective authentication mechanisms were not used to secure this procedure. Mobile IP therefore provides for data origin authentication based on MD5 hash values and time stamps (and optionally random numbers) between the different entities MN, HA and FA (see below).

❑ The use of encryption is recommended to protect the confidentiality of transmitted data although no particular scheme is specified for this purpose.

❑ The specification notes that the confidentiality of the current location of a mobile node can be protected through the use of reverse tunnels.

❑ It is remarked that appropriate key management is an important prerequisite for establishing necessary authentication relationships. The standard, however, does not discuss this requirement further.

Authentication with Mobile IP

The authentication of the registration procedure involves 'attaching' message authentication codes (MACs) to registration messages. Mobile IP provides for a range of *extensions* for this purpose: *mobile-home authentication extension* for the mutual authentication of mobile node and home agent, *mobile-foreign authentication extension* for the authentication relationship between mobile node and foreign agent and *foreign-home authentication extension* for the authentication relationship between foreign and home agent.

Only the authentication between mobile node and home agent is compulsory; the other two authentication relationships listed above are optional. According to RFC 2002 (and RFC 3220), both can be excluded if appropriate key management is not available and these relationships cannot be established. This circumstance is taken into account in the following sequence where the respective authentication values denoted with $Sig_{MN,FA}$, $Sig_{FA,HA}$, etc. are placed in square brackets '[ ]'. An exchange of the following messages signals the registration of a mobile node:

1. $MN \to FA:$ $\big(RegReq, Flags, Lifetime, Addr_{MN}, Addr_{HA}, COA,$
$Id_{Req}, NAI_{MN}, Sig_{MN,HA}[, Sig_{MN,FA}]\big)$

2. $FA \to HA:$ $\big(RegReq, Flags, Lifetime, Addr_{MN}, Addr_{HA}, COA,$
$Id_{Req}, NAI_{MN}, Sig_{MN,HA}[, Sig_{MN,FA}][, Sig_{FA,HA}]\big)$

3. $HA \to FA:$ $\big(RegRep, Code, Lifetime, Addr_{MN}, Addr_{HA},$
$Id_{Rep}, NAI_{MN}, Sig_{HA,MN}[, Sig_{HA,FA}]\big)$

4. $FA \to MN:$ $\big(RegRep, Code, Lifetime, Addr_{MN}, Addr_{HA},$
$Id_{Rep}, NAI_{MN}, Sig_{HA,MN}[, Sig_{HA,FA}][, Sig_{FA,MN}]\big)$

In the messages above $RegReq$ and $RegRep$ signify a registration request or reply. The field $Flags$ is used to request a range of options that will not be discussed here. The field $Lifetime$ indicates the desired or granted period of validity of the registration, and $Addr_{MN}$ and $Addr_{HA}$ give the IP address of the mobile node and its home agent. The field COA shows the desired care-of-address for the mobile node and the fields Id_{Req} and Id_{Rep} are used simultaneously to allocate responses to requests and to protect registration messages from being replayed as they contain time stamps.

Network access identifier (NAI)

The field NAI_{MN} represents an extension of the Mobile IP standard specified in RFC 2794 [CP00]. It identifies mobile nodes when no globally valid home-IP address is used for this purpose and will be discussed in detail in the following section.

Computation of MACs

The MACs $Sig_{MN,HA}$, $Sig_{MN,FA}$, etc., are always computed over the entire registration message, including all the extensions that occur before the respective value. The original Mobile IP standard RFC 2002 [Per96] specifies the use of prefix-suffix mode (see also Section 5.4) for this purpose. Due to cryptographic concerns, the updated standard RFC 3220 [Per02] specifies deployment of the now common HMAC construction instead [KBC97].

Authentication relationships

The description of Mobile IP registration indicates that the standard provides for three different authentication relationships only one of which, that between mobile node and its home agent, is specified as compulsory. A fundamental question in this context regards the motivations that can be identified for the separate authentication relationships between a mobile node, the access network the node is currently using and its home network.

❏ *Mobile node ↔ home network*: The main purpose of this relationship is to prevent attackers from re-directing data packets addressed to the mobile node to a different destination address through execution of the registration procedure (*'hijacking attack'*).

❑ *Mobile node ↔ foreign network*: This authentication relationship provides the basis for controlling which mobile nodes are permitted to use the resources of the access network. It is also an important prerequisite for verifiable *'accounting'* of the resources used by individual mobile nodes.

❑ *Foreign network ↔ home network*: In principle, this relationship pursues the same goals as authentication between mobile node and foreign network. For practical reasons it also has to be assumed that a foreign network *a priori* cannot know every single mobile node and instead establishes a relationship of trust with the mobile node through its home network. This gives the relationship to the home network an additional role, which is to provide the basis for establishing the mobile node-foreign network relationship. The relationship between foreign and home network also enables the home network to control which access networks can be used by its mobile nodes.

The original Mobile IP standard unfortunately offers no procedures for establishing an MN-FA relationship over the FA-HA relationship.

Lack of key   Particularly because of its lack of procedures for key negotia-
negotiation and   tion and distribution, Mobile IP does not have adequate system
distribution   functions to implement authentication and key management for true mobile Internet communication. Th IETF also recognised this deficiency and initiated efforts to solve the problem using a scalable infrastructure for *authentication, authorization and accounting (AAA)*, which is described in the following section.

17.3   Integrated AAA/Mobile IP Authentication

The standardisation of a scalable infrastructure for *Authentication, Authorization and Accounting (AAA)* was particularly motivated by the need of Internet service providers (ISPs) to compile and invoice the services used by customers of other ISPs and therefore, primarily addresses the requirements of fixed network operators. The solutions developed in this connection are also useful for the operation of mobile Internet communication and efforts are

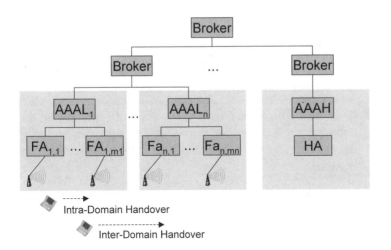

Figure 17.6
*Overview of AAA
support for Mobile IP*

underway to consider the explicit requirements of mobile communication.

Figure 17.6 presents an overview of the approach the IETF pursued to support Mobile IP using a AAA infrastructure [GHJP00]. The upper part of the illustration shows the actual AAA infrastructure consisting of AAA servers (AAAL, AAAH) and the AAA brokers that negotiate between the AAA servers. The lower part shows the connection to Mobile IP entities.

The objective is to make the AAA server of an access network (usually denoted as a local AAA server, AAAL) responsible for a range of foreign agents acting as AAA clients towards it. Normally, one (or more) local AAA servers manages an *administrative domain*, which is a network area that falls under the control of a specific operator, in respect of AAA services. This results in two fundamental forms of handover:

*Administrative
domains*

❏ An *intra-domain handover* occurs when a mobile node changes the foreign agent within an administrative domain.

*Intra-domain
handover*

❏ An *inter-domain handover* occurs when a mobile node changes to a foreign agent in a different administrative domain.

*Inter-domain
handover*

In addition to the entities that already exist with standard Mobile IP, AAA-supported Mobile IP registration and the associated authentication processes include a local AAA server (AAAL) and a home AAA server (AAAH). If a trust relationship and, consequently, an authentication relationship does not exist between the

AAAL and the AAAH, a number of AAA brokers are integrated into the process as negotiators.

Two separate aspects have to be considered in this context:

❏ *Mobile IP registration*, which mainly secures the configuration for diverting data traffic for the mobile node.

❏ *AAA authentication*, which should authenticate a mobile node for the accounting purposes and also can support key distribution for the authentication as part of Mobile IP registration.

Trust model   Figure 17.7 shows the underlying trust model for this approach. A basic distinction is made in the model between *static* (shown by arrows with continuous lines) and *dynamic trust relationships* (arrows with dotted lines).

Foreign agents within an administrative domain always have a static trust relationship with their local AAA server. This trust relationship can be secured through IPSec with the 'static nature' of this relationship given by the fact that it is preconfigured on both sides.[1]

Figure 17.7
*Trust model of
AAA/Mobile IP
authentication*

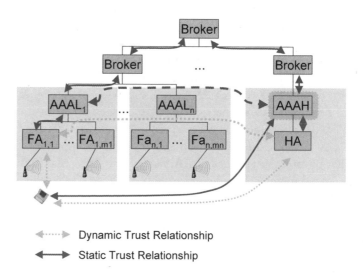

◀┈┈▶ Dynamic Trust Relationship

◀━━▶ Static Trust Relationship

*Chain of trust
between AAA servers*
    Each local AAA server additionally has a static trust relationship with at least one AAA broker, which itself also has static trust relationships with other AAA brokers. This enables a chain

---

[1]IPSec-SAs are constructed dynamically based on a security policy (see also Section 11.7). Because the policy in this case is specified statically, the trust relationship can still be seen as static.

of trust to be established between the local AAA server of a foreign network and the AAA server of the home network of a mobile node. Another alternative is for an AAAL and an AAAH to have a direct trust relationship with one another for optimisation reasons if they frequently interact with one another.

Each mobile node has a static trust relationship with its home-AAA server. This relationship enables it to authenticate itself to its AAAH from any network. The home-AAA server also maintains a static trust relationship with one or multiple Mobile IP home agents.

The static authentication relationships described form the basis for the trust relationship Mobile IP provides between mobile node, foreign agent and home agent. These relationships can be constructed dynamically using the AAA infrastructure, thereby permitting global scalability and a large degree of flexibility.

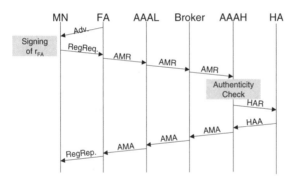

Figure 17.8
Overview of AAA/Mobile IP authentication

This idea is implemented through *integrated AAA/Mobile IP registration*. Figure 17.8 shows the message exchanges that occur as part of this registration process. The following description of the authentication protocol run during the process is based on the notation for cryptographic protocols used in this book (see also Section 7) and concentrates on the aspects that are important to authentication and key management. Consequently, unimportant message elements are not included and merely indicated by '...'.

AAA/Mobile IP registration procedure

1. All foreign agents periodically send Mobile IP-provided advertisement messages that also contain a *network access identifier (NAI) extension* [AB99, CP00] and a *challenge-response extension* [PC00]. Foreign agents use the NAI extension NAI_{FA} to identify themselves to mobile nodes. The challenge-response extension contains a fresh random num-

Advertisements of FAs

ber r_{FA} generated by the foreign agent:

$$FA \rightarrow MN: \quad \left(Advertisement, \dots, NAI_{FA}, r_{FA}\right)$$

Registration request by MN

2. The mobile node stores the NAI received from the foreign agent; creates a Mobile IP registration message, which contains the random number generated by the foreign agent, its own NAI and an HMAC-based 'signature' destined for its home-AAA server; and sends this message to the foreign agent:

$$MN \rightarrow FA: \quad \left(RegReq, \dots, r_{FA}, NAI_{MN}, Sig_{MN,AAAH}\right)$$

Generating an AAA registration request

3. Upon receipt of this message the foreign agent generates an AAA-Mobile IP registration message (AMR), which contains the Mobile IP registration message of the mobile node, and sends it to its local AAA server:

$$FA \rightarrow AAAL: \quad \left(AMR, \dots, RegReq, \dots, r_{FA}, NAI_{MN},\right.$$
$$\left. Sig_{MN,AAAH}\right)$$

Forwarding AAA registration request

4. The local AAA server forwards this message either indirectly through an AAA broker or directly to the home-AAA server of the mobile node. The home-AAA server can be derived from the NAI of the mobile node (in this example it is assumed that the message is forwarded directly to the AAAH):

$$AAAL \rightarrow AAAH: \quad \left(AMR, \dots, RegReq, \dots, r_{FA}, NAI_{MN},\right.$$
$$\left. Sig_{MN,AAAH}\right)$$

Checking the AAA registration request and generating a request to the HA

5. The home-AAA server checks the 'signature' $Sig_{MN,AAAH}$ of the original registration message created by the mobile node. If this verification is successful, the home-AAA server can assume that the mobile node really once generated the message. However, it cannot reach any conclusion about the currentness of the message because it did not generate the contained random number r_{FA} itself and therefore does not know when it was generated.

The home-AAA server thus generates a *home agent registration (HAR) message*, which again contains the original registration message of the mobile node, a session key $K_{MN,HA}$ to be used between MN and HA, and a second session key $K_{FA,HA}$ to be used between FA and HA. These two session keys are encrypted with a key $K_{AAAH,HA}$ negotiated between AAAH and HA. This message also contains the session keys $K_{MN,FA}$ and $K_{MN,HA}$ needed by the mobile node. These keys are encrypted with the key $K_{MN,AAAH}$ negotiated between MN and AAAH.[2]

The home-AAA server also appends an HMAC-based signature to the HAR message and sends the resulting message to the home agent:

$$AAAH \to HA: \quad \big(HAR, \dots, RegReq, \dots, NAI_{MN},$$
$$\{K_{MN,HA}, K_{FA,HA}\}_{K_{AAAH,HA}},$$
$$\{K_{MN,FA}, K_{MN,HA}\}_{K_{MN,AAAH}}, Sig_{AAAH,HA}\big)$$

6. Upon receipt of this message the home agent verifies the signature of the home-AAA server, registers the mobile node with the care-of-address contained in the original registration message and decrypts and stores the two session keys sent to it. Then it generates a *Mobile IP registration reply* message (RegRep), which also contains the session keys provided by the AAAH as well as an HMAC-based signature created by the home agent. The RegRep message is inserted in a *home agent answer* message (HAA), which is provided with a signature for the AAAH and sent to it:

Registration of MN and generation of registration reply

$$HA \to AAAH: \quad \big(HAA, \dots, (RegRep, \dots,$$
$$\{K_{MN,FA}, K_{MN,HA}\}_{K_{MN,AAAH}},$$
$$Sig_{HA,MN}), Sig_{HA,AAAH}\big)$$

7. The home-AAA server generates an *AAA mobile registration answer* message (AMA), which contains the RegRep message generated and signed by the home agent. If registration with the home agent is successful, the home-AAA server also attaches the session keys needed by the foreign agent to the message. The session keys are encrypted beforehand with one of the keys negotiated with the AAA peer entity.[3] The resulting message is signed by the home-AAA server and sent

Preparing an AAA registration answer

---

[2]This encryption of the session key for allocation to the mobile node is expected to involve combining the two functions XOR and MD5.

[3]If the message is being sent to AAAL through a series of AAA brokers, AAAH uses key $K_{AAAH,Broker}$ instead of key $K_{AAAH,AAAL}$ and each AAA broker performs the cryptographic verification and recoding of the contained session keys according to a 'hop-by-hop' security model. To keep the example simple, we will assume the easier case of direct communication between AAAH and AAAL.

to the local AAA server of the access network being visited by the mobile node:

$$AAAH \rightarrow AAAL: \ \big(AMA, \ldots, r_{FA}, \{K_{MN,FA}, K_{FA,HA}\}_{K_{AAAH,AAAL}},$$
$$(RegRep, \ldots, \{K_{MN,FA}, K_{MN,HA}\}_{K_{MN,AAAH}},$$
$$Sig_{HA,MN}), Sig_{AAAH,AAAL}\big)$$

Processing AAA registration answer at AAA server

8. The local AAA server verifies the contained signature and decrypts and stores the contained session keys $K_{MN,FA}$ and $K_{FA,HA}$. Then AAAL encrypts these keys again using key $K_{FA,AAAL}$ so that they can be transmitted securely to the foreign agent, and sends the following message to the foreign agent:

$$AAAL \rightarrow FA: \ \big(AMA, \ldots, r_{FA}, \{K_{MN,FA}, K_{FA,HA}\}_{K_{FA,AAAL}},$$
$$(RegRep, \ldots, \{K_{MN,FA}, K_{MN,HA}\}_{K_{MN,AAAH}},$$
$$Sig_{HA,MN}), Sig_{AAAL,FA}\big)$$

Processing registration reply at FA

9. Upon receipt of this message the foreign agent verifies the contained signature and processes the AMA message. If the AMA message indicates that the registration of the mobile node was successful, the foreign agent concludes that the mobile node has correctly signed the random number r_{FA} sent by the foreign agent during the first procedure and can therefore assume that the mobile node is authentic. The foreign agent decrypts and stores the contained session keys $K_{MN,FA}$ and $K_{FA,HA}$ and then forwards the RegRep message to the mobile node:

$$FA \rightarrow MN: \ \big(RegRep, \ldots, \{K_{MN,FA}, K_{MN,HA}\}_{K_{MN,AAAH}},$$
$$Sig_{HA,MN}\big)$$

Processing registration reply at MN

10. The mobile node decrypts and stores the contained session keys $K_{MN,FA}$ and $K_{MN,HA}$. Using key $K_{MN,HA}$, it verifies the signature $Sig_{HA,MN}$ generated by the home agent. If this verification is positive, the registration of the mobile node at the foreign agent is completed and the mobile node can now start using the services of the access network.

If the mobile node has to renew its Mobile IP registration at a later time (e.g. after expiration of *Mobile IP registration timeout*), it uses session keys $K_{MN,FA}$ and $K_{MN,HA}$. This avoids the need for a direct participation of the AAA infrastructure in the re-registration process.

If a local handover occurs within the same administrative domain, the existence of the local AAA server optimises the registration on the basis of negotiated session keys. The original plans of the IETF-AAA task group [CP01] provided this optimisation but it was removed again in later revisions of the specifications because the Mobile IP task group needed to clarify some of the implementation details. As the underlying idea shows how local handovers can be optimised with Mobile IP, the subject is of interest and should be explained at this juncture irrespective of its status within the IETF.

Potential optimisation of local handovers

When optimising local handovers, a mobile node basically analyses the network access identifier (NAI) inserted by the foreign agent into its advertisement message. If the suffix (the part after the '@' character) matches the suffix of the NAI of the previous foreign agent, the mobile node assumes that it is executing an intra-domain handover. It therefore signs the random number r_{FAnew} generated by the new foreign agent $FAnew$ with the key $K_{MN,FAold}$ and conveys the NAI of the old foreign agent $FAold$ to the new foreign agent by attaching an appropriate NAI extension to the RegReq message. The authentication procedure in this case involves the following steps:[4]

Detecting intra-domain handovers

1. Like every foreign agent, $FAnew$ periodically sends advertisement messages with its NAI and a freshly generated random number:

   Advertisement of FAnew

   $$FA \rightarrow MN: \ \left(Advertisement, \ldots, NAI_{FAnew}, r_{FAnew}\right)$$

2. Upon receipt of this message, the mobile node generates a Mobile IP RegReq message, which contains the following: a random number generated by the foreign agent, its own NAI as well as the NAI of the previous foreign agent, a signature for the home agent as well as a signature for the new foreign agent. The latter signature is prepared with the key sent to the mobile node during the AAA authentication procedure for authentication to the previous foreign agent. The resulting message is sent by the mobile node to the new foreign agent:

   Registration request of MN

   $$MN \rightarrow FA: \ \big(RegReq, \ldots, r_{FAnew}, NAI_{MN}, NAI_{FAold},$$
   $$Sig_{MN,HA}, Sig_{MN,FAold}\big)$$

---

[4]Complete specifications for exact procedures and message formats do not exist in [CP01]; therefore, only those message elements that are relevant to the underlying idea of the authentication protocol are shown here.

Generating an AAA registration request

3. The foreign agent subsequently generates an AAA mobile registration request message, which contains the Mobile IP RegReq message, and sends it to its local AAA server:

$$FA \rightarrow AAAL: \ \left(AMR, \ldots, RegReq, \ldots, r_{FAnew}, NAI_{MN}, \right.$$
$$\left. NAI_{FAold}, Sig_{MN,HA}, Sig_{MN,FAold} \right)$$

Sending session keys and preparing a response

4. The local AAA server verifies whether it can send sessions keys $K_{MN,FAold}$ and $K_{FAold,HA}$ to the new foreign agent. If the verification is positive, it updates the data sets it keeps for the mobile node and responds to the new foreign agent with a corresponding message:

$$AAAL \rightarrow FA: \ \left(AMA, \ldots, r_{FAnew}, \right.$$
$$\left\{ K_{MN,FAold}, K_{FAold,HA} \right\}_{K_{FAnew,AAAL}},$$
$$\left. Sig_{AAAL,FAnew} \right)$$

Receipt of session keys

5. Upon receipt of this message, the new foreign agent can decrypt the contained session keys, and using key $K_{MN,FAold}$, verify the signature $Sig_{MN,FAold}$ of the RegReq message. If this verification is successful, the foreign agent continues with the conventional Mobile IP registration procedure described in Section 17.2.

This authentication scheme enables intra-domain handovers to be processed efficiently because it eliminates the complete AAA authentication process required for intra-domain handovers if the local AAA server still has valid session keys for the mobile node. A complete AAA authentication procedure is only required if the session keys are no longer valid or an inter-domain handover is being executed.

Security aspects

The following comments about the described procedures can be stated from a security point of view:

Complex security analysis

❑ The authentication procedures include a range of different entities, which complicates the security analysis.

Verification of currentness is distributed

❑ The challenge-response verification is implemented on a distributed basis: The foreign agent generates a random number but cannot verify the authenticity of the response. To verify the authenticity of the response, it has to trust an authentication server that it does not know directly but only through a chain of trust. On the other side, the home-AAA server is able to verify the authenticity of a message but cannot infer anything about its currentness since it did not generate the contained random number itself.

❑ The negotiating AAA brokers can read the session keys sent to the local AAA server and the foreign agent because they always perform a cryptographic recoding ('hop-by-hop security').

'Hop-by-hop security'

❑ The NAI extension of the mobile node is transmitted unprotected in plaintext in the fixed network as well as on the radio link. Mobile IP/AAA thus offers no protection of location information for mobile nodes during registration.

Lack of anonymity

17.4   Summary

Mobile IP is currently the approach the IETF is pursuing to enable mobile devices to use Internet services. With this approach, a mobile node registers itself in its home network so that IP packets arriving for it in the home network can be forwarded to its current location (determined by its current care-of-address). The packets sent by the mobile node either contain its permanent IP address in the home network as the source address, which is usually topologically not correct in the visited network, or are returned through a reverse tunnel to the home network and then sent from the home network to the actual receiver with the mobile node's permanent IP address in the source address field.

The indirect forwarding of packets destined for a mobile node impacts efficiency and creates a higher network load. Consequently, optional route optimisation was specified, which allows a corresponding node to encapsulate its packets itself and send them directly to the current care-of-address of the mobile node. One downside to this option is that the corresponding node must implement basic mobility support functions. Another downside is the need for complicated signalling procedures each time the mobile node changes its foreign agent. This is because every corresponding node must be informed of the new care-of-address when a handover occurs for a mobile node.

Efficiency problems and network load

Route optimisation

It is important that Mobile IP registration procedures cannot be misused by attackers to re-direct packets destined for a mobile node ('traffic hijacking'). Consequently, Mobile IP provides for basic authentication of registration messages by adding message authentication codes computed with the HMAC construction. Furthermore, registration messages can optionally also be authenticated between mobile node and foreign agent as well as between foreign agent and home agent. However, Mobile IP does not pro-

Security of registration procedure

Lack of key management with standard Mobile IP

vide key management for the authentication keys needed for this purpose.

Integrated AAA/Mobile IP registration

Not least because of the last reason mentioned, use of integrated AAA/Mobile IP registration procedures, also standardised by the IETF, is recommended. These procedures permit integrated AAA authentication and Mobile IP registration using a single message traversal over the Internet (i.e. only one message traversal between the foreign network and the home network plus the processing times in the nodes passed). At the same time the registration procedures allocate the session keys needed for Mobile IP. As long as the session keys are still valid, necessary re-registrations can be performed with the Mobile IP procedures without including the AAA infrastructure. In the case of local handovers within an administrative domain, in principle the already established sessions keys can be securely communicated to the new foreign agent, which helps to avoid the relatively time-consuming process that involves a complete AAA authentication including the home network.

Lack of protection in preparation of mobility profiles with route optimisation

A serious problem with Mobile IP with route optimisation is that the current location area of a mobile node cannot be hidden from corresponding nodes if communication is being optimised with these nodes. Furthermore, binding-update messages sent to corresponding nodes are not authenticated, which means that potential attackers can maliciously divert the corresponding tunnels. In contrast to GSM and UMTS, where the movement of mobile devices can at least be hidden on the radio interface (from passive attackers) and from communication partners, Mobile IP turns out to be a considerable deterioration in this respect. This also applies to Mobile IP with authentication using an AAA infrastructure.

Secure mobile Internet communication continues to be an active research area, and this introductory chapter can only provide an initial insight into some of the problems that exist. Interested readers are therefore advised to refer to the appropriate Web pages to follow current developments in the IETF and to study relevant professional articles and publications on the topic. For example, [PSS01] analyses the security requirements of Mobile IP in a variety of deployment scenarios and derives the appropriate packet filtering and tunnel configurations appropriate for these scenarios.

17.5 Supplemental Reading

[AB99]   ABOBA, B.; BEADLES, M.:   *The Network Access Identifier*. January 1999. – RFC 2486, IETF, Status: Proposed standard, `ftp://ftp.internic.net/rfc/rfc2486.txt`

[CP00]   CALHOUN, P.; PERKINS, C.: *Mobile IP Network Access Identifier Extension for IPv4*. March 2000. – RFC 2794, IETF, Status: Proposed standard, `ftp://ftp.internic.net/rfc/rfc2794.txt`

[CP01]   CALHOUN, P. R.; PERKINS, C. E.:   *Diameter Mobile IP Extensions*.   Internet draft (work in progress).   February 2001. –   `http://search.ietf.org/internet-drafts/draft-calhoun-diameter-mobileip-%12.txt`

[GHJP00] GLASS, S.; HILLER, T.; JACOBS, S.; PERKINS, C.: *Mobile IP Authentication, Authorization, and Accounting Requirements*. October 2000. – RFC 2977, IETF, Status: Informational, `ftp://ftp.internic.net/rfc/rfc2977.txt`

[Mon98]   MONTENEGRO, G.: *Reverse Tunneling for Mobile IP*. May 1998. – RFC 2344, IETF, Status: Proposed standard, made obsolete by RFC 3024, `ftp://ftp.internic.net/rfc/rfc2344.txt`

[Mon01]   MONTENEGRO, G.: *Reverse Tunneling for Mobile IP, revised*. January 2001. – RFC 3024, IETF, Status: Proposed standard, obsoletes RFC 2344, `ftp://ftp.internic.net/rfc/rfc3024.txt`

[PC00]   PERKINS, C.; CALHOUN, P.:   *Mobile IPv4 Challenge/Response Extensions*. November 2000. – RFC 3012, IETF, Status: Proposed standard, `ftp://ftp.internic.net/rfc/rfc3012.txt`

[Per96]   PERKINS, C.: *IP Mobility Support*. October 1996. – RFC 2002, IETF, Status: Proposed standard, made obsolete by RFC 3220, `ftp://ftp.internic.net/rfc/rfc2002.txt`

[Per02]   PERKINS, C.: *IP Mobility Support, Revised*. January 2002. – RFC 3220, IETF, Status: Proposed stan-

dard, obsoletes RFC 2002, ftp://ftp.internic.
net/rfc/rfc3220.txt

[PJ00]   PERKINS, C.; JOHNSON, D.B.: *Route Optimization in
           Mobile IP.* Internet draft (work in progress). February
           2000. – http://www.ietf.org/internet-drafts/
           draft-ietf-mobileip-optim-09.txt

[PSS01]   PÄHLKE, F.; SCHÄFER, G.; SCHILLER, J.:   Multi-
           lateral sichere Mobilitätsunterstützung für IP-Netze:
           Paketfilter- und Tunnelkonfiguration (Multilateral Se-
           cure Mobility Support for IP Networks: Packet Filter
           and Tunnel Configuration). In: *Praxis der Informa-
           tionsverarbeitung und Kommunikation (PIK)* 24 Octo-
           ber 2001

17.6   Questions

1.   Which security goals were considered in the standardisation
     of the original Mobile IP?

2.   Develop solutions to implement trusted communication rela-
     tionships between CNs and MNs.

3.   What security risks result from the fact that according to the
     original Mobile IP standard mobile nodes in a visited network
     use their permanent home address as the source address in
     the packets they send?

4.   Consider how a hijacking attack takes place on tunnelled traf-
     fic between a CN and an MN in the case of route optimisation.
     Which traffic direction can you intercept this way?

5.   Consider a simple attack that allows you to monitor the cur-
     rent care-of-address of a mobile node with controllable time
     resolution when route optimisation is used.

6.   Which security goals are pursued with integrated
     AAA/Mobile IP authentication?

7.   Develop an alternative for sending a network access identifier
     in plaintext with each registration over the radio interface.

8.   What is 'hop-by-hop security'? What are the risks associated
     with this security model?

9. For Mobile IP registration separate security parameter indices (SPI) are used to identify the security associations that exist between a mobile node, the foreign agent used by it and the home agent of the mobile node. These SPIs are determined by the destination node of the registration messages being secured. What are the resulting implications in the context of optimised authentication procedures for intra-domain handovers?

Bibliography

[3GP02]   3GPP: *3G Security: Security Architecture (Release 1999)*. 3rd Generation Partnership Project, Technical Specification Group Services and System Aspects, 3GPP TS 33.102, V3.12.0. June 2002

[AB99]   ABOBA, B.; BEADLES, M.: *The Network Access Identifier*. January 1999. – RFC 2486, IETF, Status: Proposed Standard, `ftp://ftp.internic.net/rfc/rfc2486.txt`

[Agn88]   AGNEW, G. B.: Random Sources for Cryptographic Systems. In: *Advances in Cryptology — Eurocrypt '87 Proceedings*, Springer, 1988, pp. 77–81

[Amo94]   AMOROSI, E. G.: *Fundamentals of Computer Security Technology*. Prentice Hall, 1994

[AS99]   ABOBA, B.; SIMON, D.: *PPP EAP TLS Authentication Protocol*. October 1999. – RFC 2716, IETF, Status: Experimental, `ftp://ftp.internic.net/rfc/rfc2716.txt`

[AT&86]   AT&T: *T7001 Random Number Generator*. Data Sheet. August 1986

[ATM97a]   ATM Forum: *Phase I ATM Security Specification*. April 1997. – (Draft Version 1.02)

[ATM97b]   ATM Forum Technical Committee: *BTD-SIG-SEC-01.00: UNI 4.0 Security Addendum*. February 1997. – ATM Forum/97-0019

[ATM99]   ATM FORUM: *ATM Security Specification Version 1.0*. February 1999. – AF-SEC- 0100.000

[BAN90]   BURROWS, M.; ABADI, M.; NEEDHAM, R.: A Logic of Authentication. In: *ACM Transactions on Computer Systems* 8, February 1990, No. 1, pp. 18–36

[BGW01]   BORISOV, N.; GOLDBERG, I.; WAGNER, D.: Intercepting Mobile Communications: The Insecurity of 802.11. In: *Proceedings of ACM MobiCom*, 2001. – http://www.cs.berkeley.edu/~daw/papers/wep-mob01.ps

[Bie90]   BIEBER, P.: A Logic of Communication in a Hostile Environment. In: *Proceedings of the Computer Security Foundations Workshop III*, IEEE Computer Society Press, June 1990, pp. 14–22

[BKS98]   BLEICHENBACHER, D.; KALISKI, B.; STADDON, J.: *Recent Results on PKCS #1: RSA Encryption Standard*. 1998. – RSA Laboratories' Bulletin 7

[BKY93]   BETH, T.; KLEIN, B.; YAHALOM, R.: Trust Relationships in Secure Systems: A Distributed Authentication Perspective. In: *Proceedings of the 1993 Symposium on Security and Privacy*, IEEE Computer Society Press, May 1993, pp. 150–164

[Bre89]   BRESSOUD, D. M.: *Factorization and Primality Testing*. Springer, 1989

[Bry88]   BRYANT, R.: *Designing an Authentication System: A Dialogue in Four Scenes*. 1988. – Project Athena, Massachusetts Institute of Technology, Cambridge, USA

[BS90]   BIHAM, E.; SHAMIR, A.: Differential Cryptanalysis of DES-like Cryptosystems. In: *Journal of Cryptology* 4, 1990, No. 1, pp. 3–72

[BS93]   BIHAM, Eli; SHAMIR, Adi: *Differential Cryptanalysis of the Data Encryption Standard*. Springer, 1993

[BS02]   BLANCO RINCON, R.; SCHÄFER, G.: Securing 802.11-WLANs Using VPN Technology with Support for Nomadic Users. In: *Proceedings of WLAN Security '02*, 2002. – Paris, France

[BV98]   BLUNK, L.; VOLLBRECHT, J.: *PPP Extensible Authentication Protocol (EAP)*. March 1998. – RFC 2284, IETF, ftp://ftp.internic.net/rfc/rfc2284.txt

[CFPR96] COPPERSMITH, D.; FRANKLIN, M. K.; PATARIN, J.; REITER, M. K.: Low Exponent RSA with Related Messages. In: MAURER, U. (Ed.): *In Advances in Cryptology – Eurocrypt'96 Proceedings*, Springer, 1996. – Vol. 1070 of Lecture Notes in Computer Science

[CLR90] CORMEN, T. H.; LEISERSON, C. E.; RIVEST, R. L.: *Introduction to Algorithms*. The MIT Press, 1990

[CP00] CALHOUN, P.; PERKINS, C.: *Mobile IP Network Access Identifier Extension for IPv4*. March 2000. – RFC 2794, IETF, Status: Proposed Standard, ftp://ftp.internic.net/rfc/rfc2794.txt

[CP01] CALHOUN, P. R.; PERKINS, C.: *Diameter Mobile IP Extensions*. Internet Draft (work in progress). February 2001. – http://search.ietf.org/internet-drafts/ draft-calhoun-diameter-mobileip-%12.txt

[CU98] CRELL, B.; UHLMANN, A.: *Einführung in Grundlagen und Protokolle der Quanteninformatik*. 1998. – NTZ Preprint 33/1998, Universität Leipzig, http://www. uni-leipzig.de/~ntz/abs/abs3398.htm

[DA99] DIERKS, T.; ALLEN, C.: *The TLS Protocol Version 1.0*. January 1999. – RFC 2246, IETF, ftp://ftp. internic.net/rfc/rfc2246.txt

[DH76] DIFFIE, W.; HELLMAN, M. E.: New Directions in Cryptography. In: *Trans. IEEE Inform. Theory, IT-22*, 1976, pp. 644–654

[DP89] DAVIES, D. W.; PRICE, W. L.: *Security for Computer Networks*. John Wiley & Sons, 1989

[DS81] DENNING, D. E.; SACCO, G. M.: Timestamps in Key Distribution Protocols. In: *Communications of the ACM* 24, 1981, No. 8, pp. 198–208

[ElG85] ELGAMAL, T.: A Public Key Cryptosystem and a Signature Scheme based on Discrete Logarithms. In: *IEEE Transactions on Information Theory* 31, July 1985, No. 4, pp. 469–472

[ESSS98] ELSTNER, J.; SCHÄFER, G.; SCHILLER, J.; SEITZ, J.: A Comparison of Current Approaches to Securing ATM

Networks. In: *Proceedings of the 6th International Conference on Telecommunication Systems*, 1998. – Nashville, TN, USA, pp. 407–415

[Fer02]   FERGUSON, N.: *IEEE P802.11 Wireless LANs – Michael: an improved MIC for 802.11 WEP.* January 2002. – Institute of Electrical and Electronics Engineers (IEEE), Document IEEE 802.11-02/020r0

[FH98]   FERGUSON, P.; HUSTON, G.: *What is a VPN?* 1998. – The Internet Protocol Journal, Volume 1, No. 1 and 2, Cisco Systems

[FKK96]   FREIER, A. O.; KARLTON, P.; KOCHER, P. C.: *The SSL Protocol Version 3.0.* 1996. – Netscape Communications Corporation

[FMS01]   FLUHRER, S.; MANTIN, I.; SHAMIR, A.: Weaknesses in the Key Scheduling Algorithm of RC4. In: *Selected Areas in Cryptography, Lecture Notes in Computer Science* Vol. 2259, Springer, 2001, pp. 1–24

[For94]   FORD, Warwick: *Computer Communications Security – Principles, Standard Protocols and Techniques.* Prentice Hall, 1994

[GHJP00]   GLASS, S.; HILLER, T.; JACOBS, S.; PERKINS, C.: *Mobile IP Authentication, Authorization, and Accounting Requirements.* October 2000. – RFC 2977, IETF, Status: Informational, `ftp://ftp.internic.net/rfc/rfc2977.txt`

[GNY90]   GONG, L.; NEEDHAM, R. M.; YAHALOM, R.: Reasoning about Belief in Cryptographic Protocols. In: *Symposium on Research in Security and Privacy* IEEE Computer Society, IEEE Computer Society Press, May 1990, S. 234–248

[GS91]   GAARDNER, K.; SNEKKENES, E.: Applying a Formal Analysis Technique to the CCITT X.509 Strong Two-Way Authentication Protocol. In: *Journal of Cryptology* 3, 1991, No. 2, pp. 81–98

[Gud85]   GUDE, M.: Concept for a High Performance Random Number Generator Based on Physical Random Phenomena. In: *Frequenz* 39, 1985, pp. 187–190

[Gud87] GUDE, M.: *Ein quasi-idealer Gleichverteilungsgenerator basierend auf physikalischen Zufallsphänomenen (An Almost Ideal Random Number Generator Based on Physical Random Phenomena)*. Dissertation, Universität Aachen, 1987

[HC98] HARKINS, D.; CARREL, D.: *The Internet Key Exchange (IKE)*. November 1998. – RFC 2409, IETF, Status: Proposed Standard, `ftp://ftp.internic.net/rfc/rfc2409.txt`

[HMNS98] HALLER, N.; METZ, C.; NESSER, P.; STRAW, M.: *A One-Time Password System*. February 1998. – RFC 2289, IETF, Status: Draft Standard, `ftp://ftp.internic.net/rfc/rfc2289.txt`

[HPV⁺99] HAMZEH, K.; PALL, G.; VERTHEIN, W.; TAARUD, J.; LITTLE, W.; ZORN, G.: *Point-to-Point Tunneling Protocol*. July 1999. – RFC 2637, IETF, Status: Informational, `ftp://ftp.internic.net/rfc/rfc2637.txt`

[HW01] HOUSLEY, R.; WHITING, D.: *IEEE P802.11 Wireless LANs – Temporal Key Hash*. December 2001. – Institute of Electrical and Electronics Engineers (IEEE), Document IEEE 802.11-01/550r3

[IEE97] INSTITUTE OF ELECTRICAL AND ELECTRONICS ENGINEERS (IEEE): *Wireless LAN Medium Access Control (MAC) and Physical Layer (PHY) Specifications*. 1997. – The Institute of Electrical and Electronics Engineers (IEEE), IEEE Std 802.11-1997

[IEE01] INSTITUTE OF ELECTRICAL AND ELECTRONICS ENGINEERS (IEEE): *Standards for Local and Metropolitan Area Networks: Standard for Port Based Network Access Control*. 2001. – IEEE Draft P802.1X/D11

[ITU87] ITU-T: *Draft Recommendation X.509: The Directory Authentication Framework, Version 7*. November 1987

[ITU93] ITU-T: *X.509: Information Technology – Open Systems Interconnection – The Directory: Authentication Framework (4)*. 1993

[KA98a]   KENT, S.; ATKINSON, R.: *IP Authentication Header.*
           November 1998. – RFC 2402, IETF, Status: Pro-
           posed Standard, `ftp://ftp.internic.net/rfc/`
           `rfc2402.txt`

[KA98b]   KENT, S.; ATKINSON, R.: *IP Encapsulating Security*
           Payload (ESP). November 1998. – RFC 2406, IETF,
           Status: Proposed Standard, `ftp://ftp.internic.`
           `net/rfc/rfc2406.txt`

[KA98c]   KENT, S.; ATKINSON, R.: *Security Architecture for the*
           Internet Protocol. November 1998. – RFC 2401, IETF,
           Status: Proposed Standard, `ftp://ftp.internic.`
           `net/rfc/rfc2401.txt`

[KBC97]   KRAWCZYK, H.; BELLARE, M.; CANETTI, R.: *HMAC:*
           Keyed-Hashing for Message Authentication, February
           1997. – RFC 2104

[Kem89]   KEMMERER, R. A.: Analyzing Encryption Protocols
           using Formal Description Techniques. In: *IEEE Jour-*
           nal on Selected Areas in Communications 7, May
           1989, No. 4, pp. 488–457

[Ker83]   KERCKHOFF, A.: La Cryptographie Militaire. In:
           Journal des Sciences Militaires, January 1883

[KNT94]   KOHL, J.; NEUMAN, B.; TS'O, T.: The Evolution of
           the Kerberos Authentication Service. In: BRAZIER,
           F.; JOHANSEN, D. (Eds): *Distributed Open Systems*,
           IEEE Computer Society Press, 1994

[Kob87]   KOBLITZ, N.: *A Course in Number Theory and Cryp-*
           tography. Springer, 1987

[Koh89]   KOHL, J.: The Use of Encryption in Kerberos for Net-
           work Authentication. In: *Proceedings of Crypto'89*,
           Springer, 1989

[KP00]   KEROMYTIS, A.; PROVOS, N.: *The Use of HMAC-*
           RIPEMD-160-96 within ESP and AH. June 2000. –
           RFC 2857, IETF, Status: Proposed Standard, `ftp:`
           `//ftp.internic.net/rfc/rfc2857.txt`

[Kum98]   KUMMERT, H.: *The PPP Triple-DES Encryption Pro-*
           tocol (3DESE). September 1998. – RFC 2420, IETF,

Status: Proposed Standard, `ftp://ftp.internic.net/rfc/rfc2420.txt`

[KW94]   KESSLER, V.; WEDEL, G: AUTOLOG – An Advanced Logic of Authentication. In: *Proceedings of the Computer Security Foundations Workshop VII*, IEEE Computer Society Press, 1994, pp. 90–99

[Les03]   LESCUYER, P.: *UMTS: Its Origins, Architecture and the Standard.* Springer, 2003

[LR92]   LONGLEY, D.; RIGBY, S.: An Automatic Search for Security Flaws in Key Management Schemes. In: *Computers & Security* 11 (1992), No. 1, pp. 75–89

[LS92]   LLOYD, B.; SIMPSON, W.: *PPP Authentication Protocols.* October 1992. – RFC 1334, IETF, Status: Obsoleted by RFC1994, `ftp://ftp.internic.net/rfc/rfc1334.txt`

[Mat94]   MATSUI, M.: Linear Cryptanalysis Method for DES Cipher. In: *Advances in Cryptology – EuroCrypt '93 Proceedings*, Springer, 1994, pp. 386–397

[MB93]   MAO, W.; BOYD, C.: Towards Formal Analysis of Security Protocols. In: *Proceedings of the Computer Security Foundations Workshop VI*, IEEE Computer Society Press, 1993, pp. 147–158

[MCF87]   MILLEN, J. K.; CLARK, S. C.; FREEDMAN, S. B.: The Interrogator: Protocol Security Analysis. In: *IEEE Transactions on Software Engineering* 13, February 1987, No. 2, pp. 274–288

[MD98]   MADSON, C.; DORASWAMY, N.: *The ESP DES-CBC Cipher Algorithm With Explicit IV.* November 1998. – RFC 2405, IETF, Status: Proposed Standard, `ftp://ftp.internic.net/rfc/rfc2405.txt`

[Mea92]   MEADOWS, C.: Applying Formal Methods to the Analysis of a Key Management Protocol. In: *Journal of Computer Security*, 1992, No. 1, pp. 5–35

[Mea95]   MEADOWS, C.: Formal Verification of Cryptographic Protocols: A Survey. In: *Advances in Cryptology – Asiacrypt '94*, Springer, 1995 (Lecture Notes in Computer Science 917), pp. 133–150

[Men93]   MENEZES, A. J.: *Elliptic Curve Public Key Cryptosystems*. Kluwer Academic Publishers, 1993

[Mer83]   MERRIT, M.: *Cryptographic Protocols*. PhD Thesis, Georgia Institute of Technology, GIT-ICS-83, February 1983

[Mer89]   MERKLE, R.: One Way Hash Functions and DES. In: *Proceedings of Crypto '89*, Springer, 1989

[Mey96]   MEYER, G.: *The PPP Encryption Control Protocol (ECP)*. June 1996. – RFC 1968 , IETF, Status: Proposed Standard, `ftp://ftp.internic.net/rfc/rfc1968.txt`

[MG98a]   MADSON, C.; GLENN, R.: *The Use of HMAC-MD5-96 within ESP and AH*. November 1998. – RFC 2403, IETF, Status: Proposed Standard, `ftp://ftp.internic.net/rfc/rfc2403.txt`

[MG98b]   MADSON, C.; GLENN, R.: *The Use of HMAC-SHA-1-96 within ESP and AH*. November 1998. – RFC 2404, IETF, Status: Proposed Standard, `ftp://ftp.internic.net/rfc/rfc2404.txt`

[MM78]   MATYAS, S. M.; MEYER, C. H.: Generation, Distribution and Installation of Cryptographic Keys. In: *IBM Systems Journal* 17, May 1978, No. 2, pp. 126–137

[Mon98]   MONTENEGRO, G.: *Reverse Tunneling for Mobile IP*. May 1998. – RFC 2344, IETF, Status: Proposed Standard, obsoleted by RFC 3024, `ftp://ftp.internic.net/rfc/rfc2344.txt`

[Mon01]   MONTENEGRO, G.: *Reverse Tunneling for Mobile IP, revised*. January 2001. – RFC 3024, IETF, Status: Proposed Standard, obsoletes RFC 2344, `ftp://ftp.internic.net/rfc/rfc3024.txt`

[Mos89]   MOSER, L.: A Logic of Knowledge and Belief for Reasoning about Computer Security. In: *Proceedings of the Computer Security Foundations Workshop II*, IEEE Computer Society Press, June 1989, pp. 57–63

[MOV97]   MENEZES, A.; VAN OORSCHOT, P.; VANSTONE, S.: *Handbook of Applied Cryptography*. CRC Press LLC, 1997

[MR99]   MÜLLER, G.; RANNENBERG, K. (Eds): *Multilateral Security in Communications*. Addison-Wesley-Longman, 1999

[MSST98]   MAUGHAN, D.; SCHERTLER, M.; SCHNEIDER, M.; TURNER, J.: *Internet Security Association and Key Management Protocol (ISAKMP)*. November 1998. – RFC 2408, IETF, Status: Proposed Standard, `ftp://ftp.internic.net/rfc/rfc2408.txt`

[NIS77]   NIST (NATIONAL INSTITUTE OF STANDARDS AND TECHNOLOGY): *FIPS (Federal Information Processing Standard) Publication 46: Data Encryption Standard*. 1977

[NIS88]   NIST (NATIONAL INSTITUTE OF STANDARDS AND TECHNOLOGY): *FIPS (Federal Information Processing Standard) Publication 46-1: Data Encryption Standard*. 1988. – updates FIPS Publication 46

[NIS01]   NIST (NATIONAL INSTITUTE OF STANDARDS AND TECHNOLOGY): *FIPS (Federal Information Processing Standard) Publication 197: Specification for the Advanced Encryption Standard (AES)*. 2001

[NS78]   NEEDHAM, R. M.; SCHROEDER, M. D.: Using Encryption for Authentication in Large Networks of Computers. In: *Communications of the ACM* 21, December 1978, No. 12, pp. 993–999

[NS87]   NEEDHAM, R.; SCHROEDER, M.: Authentication Revisited. In: *Operating Systems Review* 21, 1987, No. 1

[NZ80]   NIVEN, I.; ZUCKERMAN, H.: *An Introduction to the Theory of Numbers*. 4th edition, John Wiley & Sons, 1980

[Oor93]   VAN OORSCHOT, P. C.: Extending Cryptographic Logics of Belief to Key Agreement Protocols. In: ASHBY, V. (Ed.): *1st ACM Conference on Computer and Communications Security*. Fairfax, Virginia, ACM Press, November 1993, pp. 232–243

[OR87]   OTWAY, D.; REES, O.: Efficient and Timely Mutual Authentication. In: *Operating Systems Review* 21, 1987, No. 1

[PA98]   PEREIRA, R.; ADAMS, R.: *The ESP CBC-Mode Cipher Algorithms*. November 1998. – RFC 2451, IETF, Status: Proposed Standard, `ftp://ftp.internic.net/rfc/rfc2451.txt`

[PC00]   PERKINS, C.; CALHOUN, P.: *Mobile IPv4 Challenge/Response Extensions*. November 2000. – RFC 3012, IETF, Status: Proposed Standard, `ftp://ftp.internic.net/rfc/rfc3012.txt`

[PD00]   PETERSON, L.; DAVIE, B.: *Computernetze – Ein modernes Lehrbuch*. dpunkt.verlag, 2000

[Per96]   PERKINS, C.: *IP Mobility Support*. October 1996. – RFC 2002, IETF, Status: Proposed Standard, obsoleted by RFC 3220, `ftp://ftp.internic.net/rfc/rfc2002.txt`

[Per02]   PERKINS, C.: *IP Mobility Support, revised*. January 2002. – RFC 3220, IETF, Status: Proposed Standard, obsoletes RFC 2002, `ftp://ftp.internic.net/rfc/rfc3220.txt`

[Pip98]   PIPER, D.: *The Internet IP Security Domain of Interpretation for ISAKMP*. November 1998. – RFC 2407, IETF, Status: Proposed Standard, `ftp://ftp.internic.net/rfc/rfc2407.txt`

[PJ00]   PERKINS, E.C.; JOHNSON, D.B.: *Route Optimization in Mobile IP*. Internet Draft (work in progress). February 2000. – `http://www.ietf.org/internet-drafts/draft-ietf-mobileip-optim-09.txt`

[PSS82]   PURDY, G. B.; SIMMONS, G. J.; STUDIER, J. A.: A Software Protection Scheme. In: *Proceedings of the 1982 Symposium on Security and Privacy*, IEEE Computer Society Press, April 1982, pp. 99–103

[PSS01]   PÄHLKE, F.; SCHÄFER, G.; SCHILLER, J.: Multilateral sichere Mobilitätsunterstützung für IP-Netze: Paketfilter- und Tunnelkonfiguration (Multilateral Secure Mobility Support for IP Networks: Packet Filter and Tunnel Configuration). In: *Praxis der Informationsverarbeitung und Kommunikation (PIK)* 24, October 2001

[PZ01]   PALL, G.; ZORN, G.: *Microsoft Point-To-Point Encryption (MPPE) Protocol*. March 2001. – RFC 3078, IETF, Status: Informational, `ftp://ftp.internic.net/rfc/rfc3078.txt`

[Ran88]   RANGAN, P. V.: An axiomatic Basis of Trust in Distributed Systems. In: *Proceedings of the 1988 Symposium on Security and Privacy*, IEEE Computer Society Press, April 1988, pp. 204–211

[RE95]   RANKL, W.; EFFING, W.: *Handbuch der Chipkarten*. Hanser, 1995

[Ric92]   RICHTER, M.: *Ein Rauschgenerator zur Gewinnung von quasi-idealen Zufallszahlen für die stochastische Simulation*. Dissertation, Universität Aachen, 1992

[Riv90]   RIVEST, R.: Cryptography. In: VAN LEEUWEN, J. (Ed.): *Handbook of Theoretical Computer Science* Vol. 1, Elsevier, 1990, pp. 717–755

[Riv91]   RIVEST, R. L.: The MD4 Message Digest Algorithm. In: *Advances in Cryptology — Crypto '90 Proceedings*, Springer, 1991, pp. 303–311

[Riv92]   RIVEST, R. L.: *The MD5 Message Digest Algorithm*, April 1992. – RFC 1321

[Riv01]   RIVEST, R.: *RSA Security Response to Weaknesses in Key Scheduling Algorithm of RC4*. 2001. – `http://www.rsa.com/rsalabs/technotes/wep.html`

[Rob96]   ROBSHAW, M.: *On Recent Results for MD2, MD4 and MD5*. November 1996. – RSA Laboratories' Bulletin, No. 4

[Rom88]   ROMKEY, J. L.: *Nonstandard for transmission of IP datagrams over serial lines: SLIP*. June 1988. – RFC 1055, IETF, Status: Standard, `ftp://ftp.internic.net/rfc/rfc1055.txt`

[RSA78]   RIVEST, R.; SHAMIR, A.; ADLEMAN, L.: A Method for Obtaining Digital Signatures and Public Key Cryptosystems. In: *Communications of the ACM*, February 1978

[RWRS00] RIGNEY, C.; WILLENS, S.; RUBENS, A.; SIMPSON, W.: *Remote Authentication Dial In User Service (RADIUS)*. June 2000. – RFC 2865, IETF, Status: Draft Standard, `ftp://ftp.internic.net/rfc/rfc2865.txt`

[SCFY96] SANDHU, R.; COYNE, E.; FEINSTEIN, H.; YOUMAN, C.: Role-Based Access Control Models. In: *IEEE Computer* 29, February 1996, No. 2, pp. 38–47

[Sch96] SCHNEIER, B.: *Applied Cryptography Second Edition: Protocols, Algorithms and Source Code in C*. John Wiley & Sons, 1996

[Sch98] SCHÄFER, G.: *Effiziente Authentisierung und Schlüsselverwaltung in Hochleistungsnetzen (Efficient Authentication and Key Management in High Performance Networks)*. October 1998. – Dissertation at the Faculty of Informatics, Universität Karlsruhe (TH), Germany

[Sch03] SCHILLER, J.: *Mobile Communications*. Second edition, Pearson Education, 2003

[Sem96] SEMERIA, C.: *Internet Firewalls and Security*. 1996. – 3Com Technical Paper

[SHB95] STEVENSON, D.; HILLERY, N.; BYRD, G.: Secure Communications in ATM Networks. In: *Communications of the ACM* 38, February 1995, pp. 45–52

[Sid86] SIDHU, D. P.: Authentication Protocols for Computer Networks: I. In: *Computer Networks and ISDN Systems* 11, 1986, No. 4, pp. 297–310

[Sim85] SIMMONS, G. J.: How to (Selectively) Broadcast a Secret. In: *Proceedings of the 1985 Symposium on Security and Privacy*, IEEE Computer Society Press, April 1985, pp. 108–113

[Sim94a] SIMMONS, G. J.: Cryptology. In: *Encyclopaedia Britannica*, Britannica, 1994

[Sim94b] SIMPSON, W.: *The Point-to-Point Protocol (PPP)*. July 1994. – RFC 1661, IETF, Status: Standard, `ftp://ftp.internic.net/rfc/rfc1661.txt`

[Sim94c]   SIMPSON, W.: *PPP in HDLC-like Framing.* July 1994.
             – RFC 1662, IETF, Status: Standard, `ftp://ftp.`
             `internic.net/rfc/rfc1662.txt`

[Sim96]   SIMPSON, W.: *PPP Challenge Handshake Authen-*
             tication Protocol (CHAP). August 1996. – RFC
             1994, IETF, Status: Draft Standard, `ftp://ftp.`
             `internic.net/rfc/rfc1994.txt`

[SIR01]   STUBBLEFIELD, A.; IOANNIDIS, J.; RUBIN, A. D.: *Us-*
             ing the Fluhrer, Mantin, and Shamir Attack to Break
             WEP. August 2001. – AT&T Labs Technical Report
             TD-4ZCPZZ

[SKW$^+$98] SCHNEIER, B.; KELSEY, J.; WHITING, D.; WAGNER,
             D.; HALL, C.; FERGUSON, N.: *TwoFish: A 128-Bit*
             Block Cipher. 1998. – `http://www.counterpane.`
             `com/twofish.html`

[SM98a]   SCHNEIER, B.; MUDGE: Cryptanalysis of Microsoft's
             Point-to-Point Tunneling Protocol (PPTP). In: *ACM*
             Conference on Computer and Communications Secu-
             rity, 1998, pp. 132–141

[SM98b]   SKLOWER, K.; MEYER, G.: *The PPP DES Encryp-*
             tion Protocol, Version 2 (DESE-bis). September 1998.
             – RFC 2419, IETF, Status: Proposed Standard, `ftp:`
             `//ftp.internic.net/rfc/rfc2419.txt`

[SMPT01]   SHACHAM, A.; MONSOUR, B.; PEREIRA, R.; THOMAS,
             M.: *IP Payload Compression Protocol (IPComp).*
             September 2001. – RFC 3173, IETF, Status: Pro-
             posed Standard, `ftp://ftp.internic.net/rfc/`
             `rfc3173.txt`

[SMW99]   SCHNEIER, B.; MUDGE; WAGNER, D.: Cryptanalysis
             of Microsoft's PPTP Authentication Extensions (MS-
             CHAPv2). In: *International Exhibition and Congress*
             on Secure Networking – CQRE [Secure], 1999

[Sne91]   SNEKKENES, E.: Exploring the BAN Approach to Pro-
             tocol Analysis. In: *1991 IEEE Computer Society Sym-*
             posium on Research in Security and Privacy, 1991, pp.
             171–181

[SO94]   SYVERSON, P.; VAN OORSCHOT, P. C.: On Unifying Some Cryptographic Protocol Logics. In: *1994 IEEE Computer Society Symposium on Research in Security and Privacy*, 1994, pp. 14–28

[SR97]   SEXTON, M.; REID, A.: *Broadband Networking – ATM, SDH and SONET*. Artech House Publishers, 1997

[Sta95]   STALLINGS, W.: *ISDN and Broadband ISDN with Frame Relay and ATM*. 3rd edition, Prentice Hall, 1995

[Sta98]   STALLINGS, W.: *High-Speed Networks – TCP/IP and ATM Design Principles*. Prentice Hall, 1998

[Sti95]   STINSON, D. R.: *Cryptography: Theory and Practice (Discrete Mathematics and Its Applications)*. CRC Press, 1995

[SV01]   SAMARATI, P; DE CAPITANI DI VIMERCATI, S.: Access Control: Policies, Models, and Mechanisms. In: FOCARDI, R.; GORRIERI, R. (Eds): *Foundations of Security Analysis and Design; Lecture Notes in Computer Science* Vol. 2171, Springer, 2001, pp. 137–196

[Syv90]   SYVERSON, P.: Formal Semantics for Logics of Cryptographic Protocols. In: *Proceedings of the Computer Security Foundations Workshop III*, IEEE Computer Society Press, June 1990, pp. 32–41

[Syv91]   SYVERSON, P.: The Use of Logic in the Analysis of Cryptographic Protocols. In: *1991 IEEE Computer Society Symposium on Research in Security and Privacy*, 1991, pp. 156–170

[Syv93a]   SYVERSON, P.: Adding Time to a Logic of Authentication. In: *1st ACM Conference on Computer and Communications Security*, 1993, pp. 97–101

[Syv93b]   SYVERSON, P.: On Key Distribution Protocols for Repeated Authentication. In: *ACM Operating System Review* 4, October 1993, pp. 24–30

[TG93]   TC-GSM, ETSI: *GSM Security Aspects (GSM 02.09)*. European Telecommunications Standards Institute

(ETSI), Recommendation GSM 02.09, Version 3.1.0. June 1993

[Tou91] TOUSSAINT, M.-J.: *Verification of Cryptographic Protocols*. PhD Thesis, Université de Liège (Belgium), 1991

[Tou92a] TOUSSAINT, M.-J.: Deriving the Complete Knowledge of Participants in Cryptographic Protocols. In: *Advances in Cryptology — CRYPTO '91 Proceedings*, Springer, 1992, pp. 24–43

[Tou92b] TOUSSAINT, M.-J.: Seperating the Specification and Implementation Phases in Cryptology. In: *ESORICS '92 — Proceedings of the Second European Symposium on Research in Computer Security*, Springer, 1992, S. 77–101

[TS94] TC-SMG, ETSI: *European Digital Cellular Telecommunications System (Phase 2): Security Related Network Functions (GSM 03.20)*. European Telecommunications Standards Institute (ETSI), ETS 300 534. September 1994

[TVR+99] TOWNSLEY, W.; VALENCIA, A.; RUBENS, A.; PALL, G.; ZORN, G.; PALTER, B.: *Layer Two Tunneling Protocol (L2TP)*. August 1999. – RFC 2661, IETF, Status: Draft Standard, ftp://ftp.internic.net/rfc/rfc2661.txt

[Var89] VARADHARAJAN, V.: Verification of Network Security Protocols. In: *Computers & Security* 8 (1989), August, No. 8, pp. 693–708

[Var90] VARADHARAJAN, V.: Use of Formal Description Technique in the Specification of Authentication Protocols. In: *Computer Standards & Interfaces* 9, 1990, pp. 203–215

[VLK98] VALENCIA, A.; LITTLEWOOD, M.; KOLAR, T.: *Cisco Layer Two Forwarding (L2F) Protocol*. May 1998. – RFC 2341, IETF, Status: Historic, ftp://ftp.internic.net/rfc/rfc2341.txt

[VV96] VOLPE, F. P.; VOLPE, S.: *Chipkarten – Grundlagen, Technik, Anwendungen*. Heise, 1996

[WC95]   WACK, J. P.; CARNAHAN, L. J.: *Keeping Your Site Comfortably Secure: An Introduction to Internet Firewalls.* 1995. – NIST Special Publication 800-10

[WHF02]   WHITING, D.; HOUSLEY, R.; FERGUSON, N.: *IEEE P802.11 Wireless LANs – AES Encryption & Authentication Using CTR Mode & CBC-MAC.* May 2002. – Institute of Electrical and Electronics Engineers (IEEE), Document IEEE 802.11-02/001r2

[WL93]   WOO, T. Y. C.; LAM, S. S.: A Semantic Model for Authentication Protocols. In: *1993 IEEE Computer Society Symposium on Research in Security and Privacy,* 1993, pp. 178–194

[YKS+01a]   YLONEN, T.; KIVINEN, T.; SAARINEN, M.; RINNE, T.; LEHTINEN, S.: *SSH Authentication Protocol.* 2001. – Internet Draft (work in progress), draft-ietf-secsh-userauth-11.txt

[YKS+01b]   YLONEN, T.; KIVINEN, T.; SAARINEN, M.; RINNE, T.; LEHTINEN, S.: *SSH Connection Protocol.* 2001. – Internet Draft (work in progress), draft-ietf-secsh-connect-11.txt

[YKS+01c]   YLONEN, T.; KIVINEN, T.; SAARINEN, M.; RINNE, T.; LEHTINEN, S.: *SSH Protocol Architecture.* 2001. – Internet Draft (work in progress), draft-ietf-secsh-architecture-09.txt

[YKS+01d]   YLONEN, T.; KIVINEN, T.; SAARINEN, M.; RINNE, T.; LEHTINEN, S.: *SSH Transport Layer Protocol.* 2001. – Internet Draft (work in progress), draft-ietf-secsh-transport-09.txt

[Yuv79]   YUVAL, G.: How to Swindle Rabin. In: *Cryptologia,* July 1979

[ZC98]   ZORN, G.; COBB, S.: *Microsoft PPP CHAP Extensions.* October 1998. – RFC 2433, IETF, Status: Informational, ftp://ftp.internic.net/rfc/rfc2433.txt

[ZCC00]   ZWICKY, E.; COOPER, S.; CHAPMAN, B.: *Building Internet Firewalls.* Second edition, O'Reilly, 2000

Abbreviations

AAA	Authentication, Authorization and Accounting
AES	Advanced Encryption Standard
AH	Authentication Header
AK	Anonymity Key
AKA	Authentication and Key Agreement
AMF	Authentication Management Field
ANSI	American National Standards Institute
AP	Access Point
ARP	Address Resolution Protocol
ARPA	Advanced Research Project Agency
ASN.1	Abstract Syntax Notation 1
AUC	Authentication Centre
AUTN	Authentication Token
AV	Authentication Vector
BSC	Base Station Controller
BSS	Base Station Sub-System (in context of GSM)
BSS	Basic Service Set (in context of WLANs)
BTS	Base Tranceiver Station
BU	Binding Update
CA	Certification Authority
CBC	Cipher Block Chaining
CFB	Ciphertext Feedback
CHAP	Challenge Handshake Authentication Protocol
CK	Confidentiality Key
CN	Corresponding Node
COA	Care-of-address
CRC	Cyclic Redundancy Check

CSPRBG	Cryptographically Secure Pseudo Random Bit Generator
CV	Chaining Value
DES	Data Encryption Standard
DHCP	Dynamic Host Configuration Protocol
DMZ	Demilitarized Zone
DOI	Domain of Interpretation
DNS	Domain Name System
DOD	Department of Defense
DSSS	Direct Sequence Spread Spectrum
EAP	Extensible Authentication Protocol
EAPOL	EAP over LANs
ECB	Electronic Code Book
EIR	Equipment Identity Register
ESP	Encapsulating Security Payload
ESS	Extended Service Set
FA	Foreign Agent
FHSS	Frequency Hop Spread Spectrum
FTP	File Transfer Protocol
GCD	Greatest Common Divisor
GMSC	Gateway Mobile Switching Centre
GPRS	General Packet Radio Service
GRE	Generic Routing Encapsulation
GSM	Global System for Mobile Communication
HA	Home Agent
HDLC	High-Level Data Link Control
HE	Home Environment
HLR	Home Location Register
HMAC	Hashed Message Authentication Code
HTTP	Hypertext Transfer Protocol
ICMP	Internet Control Message Protocol
IDEA	International Data Encryption Algorithm
IEEE	Institute of Electrical and Electronics Engineers
IETF	Internet Engineering Task Force
IK	Integrity Key
IKE	Internet Key Exchange

IMSI	International Mobile Subscriber Identity
IMT	International Mobile Telecommunications
IP	Internet Protocol
IPComp	IP Payload Compression Protocol
IPv4	Internet Protocol Version 4
IPSec	IP Security Architecture
ISAKMP	Internet Security Association and Key Management Protocol
ISDN	Integrated Services Digital Network
ISO	International Organization for Standardization
ISP	Internet Service Provider
ITU	International Telecommunications Union
IV	Initialisation Vector
IWF	Interworking Function
L2F	Layer 2 Forwarding Protocol
L2TP	Layer 2 Tunneling Protocol
LAI	Location Area Identifier
LAN	Local Area Network
LPR	Line Printing Protocol
MAC	Medium Access Control
MAC	Message Authentication Code
MAN	Metropolitan Area Network
MD	Message Digest
MDC	Modification Detection Code
MIB	Management Information Base
MIPS	Million Instructions Per Second
MN	Mobile Node
MS	Mobile Station
MSC	Mobile Switching Centre
MSISDN	Mobile Subscriber International ISDN Number
NAI	Network Access Identifier
NAT	Network Address Translation
NFS	Network File System
NIST	National Institute of Standards and Technology
NSA	National Security Agency

NSS	Network Sub-System
NTP	Network Time Protocol
OAEP	Optimal Asymmetric Encryption Padding
OSI	Open Systems Interconnection
OTP	One Time Password
OFB	Output Feedback
OFDM	Orthogonal Frequency Division Multiplexing
OMC	Operation and Management Centre
OSS	Operation Sub-System
OUI	Organizational Unit Identifier
PAE	Port Access Entities
PAP	Password Authentication Protocol
PCBC	Propagating Cipher Block Chaining
PCT	Private Communication Technology
PDN	Packet Data Network
PDU	Protocol Data Unit
POP	Point of Presence
PPP	Point-to-Point Protocol
PPTP	Point-to-Point Tunneling Protocol
PRBG	Pseudo Random Bit Generator
PSTN	Public Switched Telecommunication Network
RADIUS	Remote Authentication Dial In User Service
RAND	Random Challenge
RAS	Remote Access Server
RBG	Random Bit Generator
RC4	Rivest Cipher 4
RFC	Request for Comments
RSA	Rivest, Shamir and Adleman
RSS	Radio Sub-System
SA	Security Association
SADB	Security Association Database
SAP	Service Access Point
SGSN	Serving GPRS Support Node
SHA	Secure Hash Algorithm
S-HTTP	Secure HTTP

SIM	Subscriber Identity Module
SLIP	Serial Line IP
SMTP	Simple Mail Transfer Protocol
SN	Serving Network
SNMP	Simple Network Management Protocol
SPD	Security Policy Database
SPI	Security Parameter Index
SPKI	Simple Public Key Infrastructure
SQN	Sequence Number
SSH	Secure Shell
SSL	Secure Socket Layer
TCP	Transport Control Protocol
TGS	Ticket Granting Server
TLS	Transport Layer Security
TMSI	Temporary Mobile Subscriber Identity
TTL	Time To Live
TTP	Trusted Third Party
UDP	User Datagram Protocol
UMTS	Universal Mobile Telecommunications System
USIM	User Services Identity Module
VLR	Visitor Location Register
VPN	Virtual Private Network
WEP	Wired Equivalent Privacy
WLAN	Wireless Local Area Network
WWW	World Wide Web
XOR	Exklusive-Or
XRES	Expected Response

Index